EAST WEST

" . . . some Toronto Residences."

QUEENWEST
953

East/West
A Guide to Where People Live in Downtown Toronto

Edited by Nancy Byrtus, Mark Fram, and Michael McClelland

Coach House Books

Society for the Study of Architecture in Canada
Societé des études d'architecture au Canada

Published and distributed by Coach House Books, for the Society for the Study of Architecture in Canada, June 2000

Fourth printing, June 2007

Manufactured in Canada

Canadian Cataloguing-in-Publication Data

Main entry under title:
East/West: a guide to living in downtown Toronto

ISBN: 1 55245 065 1

1. Toronto (Ont.)–Description and travel. 2. Buildings–Ontario–Toronto. 3. Architecture–Ontario–Toronto. 4. Neighborhood–Ontario–Toronto. I. Byrtus, Nancy, 1967- . II. Fram, Mark. III. McClelland, Michael, 1951-

FC3097.5.E27 2000 971.3'541 C00-931499-7
F1059.5.T686A2 2000

Coach House Books
401 Huron Street (rear) on bpNichol Lane, Toronto, Ontario M5S 2G5
www.chbooks.com / mail@chbooks.com

Society for the Study of Architecture in Canada
A voluntary, non-profit, charitable association
Box 2302, Station D, Ottawa, Ontario K1P 5W5

Funding for this publication has been provided by Parks Canada, the Toronto Community Foundation, and the Toronto Society of Architects.

Contents

The "*Locations*" maps on pages **xii**, **16**, **76**, and **147** are extensions to the table of contents. Map numbers are page numbers. The map on page **147** shows neighbourhood *boundaries*, according to the City of Toronto. The street map on page **x** shows the street names at a very small size (magnifier required). In any case, a larger-scale street map will be an invaluable aid to finding these places.

Preface

This publication was created with the goal of providing a *snapshot* of housing types and issues in Toronto in conjunction with the Society for the Study of Architecture in Canada (SSAC) conference "Fresh Perspectives on Housing" (June 7-10, 2000).

East/West: A Guide to Where People Live in Downtown Toronto does not attempt to include all sites, all voices, and all opinions, but a significant dimension of the publication has been to provide a forum for different voices that do not usually have the opportunity to be heard together. We have deliberately avoided influencing or homogenizing the responses of the authors for the sake of an overarching thesis. Instead, we have attempted to convey each author's contribution (text and illustrations), as part of the complex and often incongruous matrix of opinions that come from governments, heritage groups, builders, planners, lawyers, architects, landscape architects, and developers, as well as students and people who live in the centre of the city. The perspectives within this book, therefore, do not necessarily represent those of the editors or the SSAC.

The publication is organized around two cross-sections of the downtown core, one in the east end and one in the west. The eastern cross-section takes Parliament Street as its spine, running from Rosedale to the Harbour; the western cross-section follows Spadina Avenue and St George/Beverley streets, from Wychwood and Casa Loma to the Harbour. This sectional view of the city is not intended to be exclusive or divisive in any sense, but to provide a profile of housing where the juxtaposition of differences becomes as significant as the presentation of congruent neighbourhoods. Most sites, but not all, were selected because they fit on one of the cross-sections. Parts of these cross-sections are adaptable for self-guided walking tours. Some sites have been included that are off the sectional grid because they illustrate major trends, ideas, or issues that could not be easily treated within the framework of the downtown.

East/West: A Guide to Where People Live in Downtown Toronto is our invitation to visit, explore and reflect upon the neighbourhoods and housing sites of Toronto.

Michael McClelland, Marsha Kelmans and Nancy Byrtus

Acknowledgements

Sources and permissions

We gratefully acknowledge the City of Toronto Archives (abbreviated in the individual credits as "CTA") and the City Planning Division, Urban Development Services, City of Toronto ("CPD/CT"), for permission to reproduce photographs, perspective views generated from the City of Toronto's computer modeling, and excerpts from city maps.

Photographs supplied by the contributors have been credited wherever possible to the original photographers. In those cases where the original credit could not be confirmed by press time, the author or the author's firm is noted as the source, preceded by this symbol – •.

The plans of "existing residences" and the pictures of "some Toronto residences" are reproduced from the "Bruce Report" – *Report of the Lieutenant Governor's Committee on Housing Conditions in Toronto, 1934*. Toronto: Hunter Rose.

The definitions on page *xiv* are from the Oxford English Dictionary (Second Edition). The "coda" in the introduction is from John Sladek. "The Communicants." *The New S.F.* London: Arrow Books, 1969.

Support

We are especially grateful to the following sponsors and supporters, whose contributions have made this book possible:

Financial assistance

Society for the Study of Architecture in Canada
Toronto Community Foundation
Parks Canada
ERA Architects Inc.
Polymath&Thaumaturge Inc.
Public Work Architects Inc.
Toronto Society of Architects

Other support

For assistance with the acquisition of computer-modeling data, Robert Wright and Charles Cox, Center for Landscape Research, Faculty of Architecture, Landscape and Urban Design, University of Toronto; and Robert Glover, City of Toronto

For writing the initial letters of support for the project, Karen Black, Cathy Crowe, Catherine Nasmith, and Michael Shapcott

For their assistance in securing historic photographs, Sally Gibson, and Andrea Aitken at the City of Toronto Archives

For carefully reviewing the publication before it went to press, Robert G. Hill

WYCHWOOD PARK
CASA LOMA
DAVENPORT ROAD
BATHURST
SPADINA
ST. GEORGE
AURA LEE PARK
JEAN SIBELIUS SQUARE
WALMER ROAD CIRCLE
HURON STREET PLAYGROUND
TADDLE CREEK PARK
VARSITY STADIUM
ROYAL ONTARIO MUSEUM
PHILOSOPHER'S WALK
UNIVERSITY OF TORONTO
QUEEN'S PARK
ONTARIO PARLIAMENT BUILDINGS
WELLESLEY STREET
COLLEGE
ALEXANDRA PARK
BELLEVUE SQUARE
ART GALLERY OF ONTARIO
GRANGE PARK
CITY HALL
OSGOODE HALL
NATHAN PHILLIPS SQUARE
RICHMOND
ADELAIDE
ST. ANDREW'S PLAYGROUND
VICTORIA MEMORIAL PARK
CLARENCE SQUARE
WELLINGTON
FRONT
UNIVERSITY
UNION STATION
OLD FORT YORK
SKYDOME
CN TOWER
BREMNER BLVD
GARDINER EXPRESSWAY
HARBOURFRONT PARK
HARBOUR SQUARE PARK
LITTLE NORWAY PARK
CHANNEL
BASIN

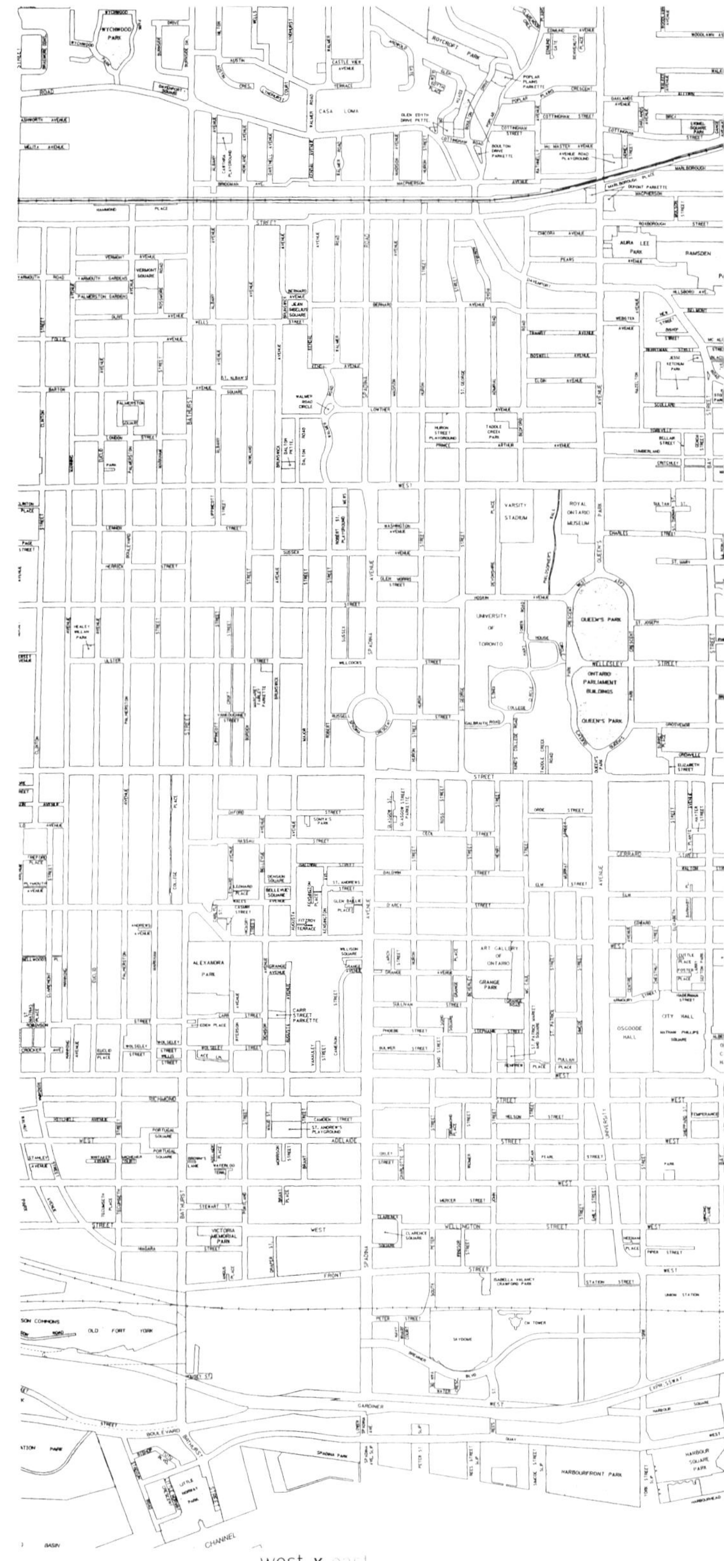

west x east

Yonge Street, though not in the precise centre of the map, is where the pages meet. The left-hand page (the verso) is the west side of the guide, and the right-hand page (the recto) is the east side of the guide. Street numbers of east-west streets begin at Yonge Street. Street numbers of north-south streets begin at the south and increase to the north.

78
DAVENPORT
80
82
DUPONT
88
84
89
AVENUE RD
90
93
91
BLOOR
92
98
99
102
100
101
HARBORD
94
97
103
104
106
105
108
107
ST GEORGE
110
112
COLLEGE
BAY
BATHURST
118
120
122
116
123
113
DUNDAS
127
126
124
BEVERLEY
128
UNIVERSITY
SPADINA
133
129
QUEEN
134
RICHMOND
132
ADELAIDE
KING
135
136
WELLINGTON
FRONT
137
GARDINER
LAKESHORE
144
140
QUEENS QUAY
138
142

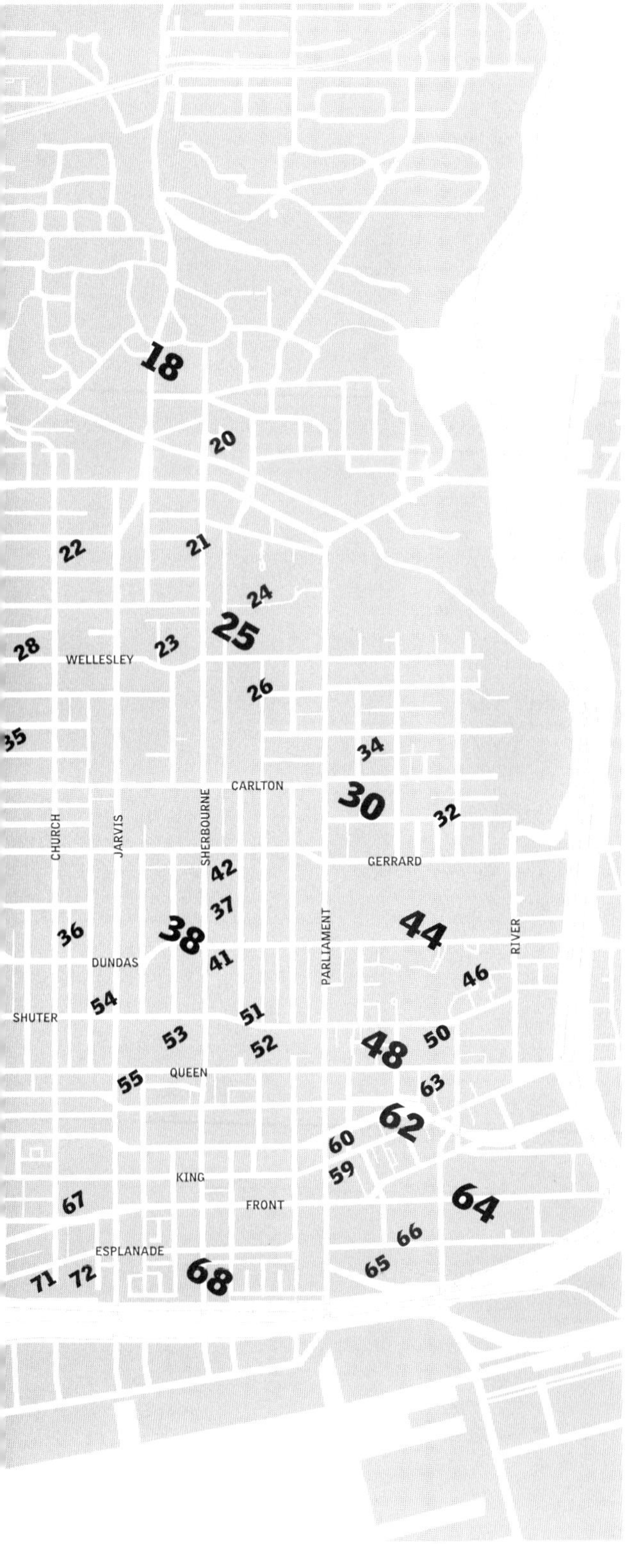

Sites
(page number)

20

Neighbourhoods
(page number)

18

18
Rosedale

25
St James Town

30
Metcalfe Street

38
Sherbourne/Dundas

44
Regent Park

48
Trefann Court

62
Corktown

64
West Don Lands

68
St Lawrence

78
Wychwood Park

84
Annex

94
Sussex-Ulster

113
Southeast Spadina

116
Spadina Avenue

137
Railway Lands

138
Central Waterfront

140
Harbourfront West

142
Bathurst Quay

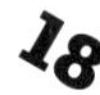

house (ha[z), *v.* [1]
[OE. *hĭsian* (in sense 1) = OHG. *h∫s n* (MHG., MLG., MDu. *husen*, Ger. *hausen*, Du. *huizen*), ON. *husa*; f. *hĭs* house *n.* [1]]
I. Transitive senses.
1. a. To receive or put into a house; to provide with a house to dwell in; to keep or store in a house or building.
b. *refl.* **To enter a house; to take refuge or shelter in a house.**
† **c. To drive or pursue into a house.** *Obs.*
2. To receive, as a house does; to give shelter to.
3. *transf.* **and** *fig.* **To place or enclose as in a house; to cover as with a roof; to harbour, lodge.**
4. a. *Naut.* **To place in a secure or unexposed position:** ***e.g.*** **a gun, by running it in on deck and fastening it by tackle, muzzle-lashing, and breeching; a topmast or topgallant-mast, by partly lowering it and fastening its heel to the mast below it.**
b. *Naut.* **To cover or protect with a roof.**
c. *Hop-growing.* **(See quot.) Cf.** housling.
d. *Carpentry.* **To fix in a socket, mortice, or the like: cf.** housing *n.* [1] **5.**
† **5. To build. (transl. L.** ***dificare.***) *Obs. rare.*
II. Intransitive senses.
† **6. To erect a house or houses; to build.** *Obs.*
7. To dwell or take shelter in (or as in) a house; to harbour. Also with ***up.***
† **8.** ***house in*** **(also in** *pass.*)**: said of a ship of which the upper works are built narrower than the lower. (Cf.** homing *vbl. n.* **1.)** *Obs.*

housing (ˈha[zɪŋ), *n.* [1]
[f. house *v.* [1] or n. [1] + -ing [1].]
1. The action of the verb house**, in various senses: †building of houses (***obs.***); putting or enclosing in a house; furnishing or provision of houses; dwelling or lodging in a house.**
2. a. Shelter of a house, or such as that of a house; house accommodation; lodging.
b. Houses or buildings collectively; house-property; *spec.* **a collection of outhouses or adjoining buildings attached to a house (dial. sometimes confused with** ***housen*****, pl. of** house**).**
c. A house or building.
† **3.** *Arch.* **A canopied niche for a statue, a tabernacle; also** *collect.* **tabernacle-work.** *Obs.*
4. *Naut.*
a. A covering or roofing for a ship when laid up, or under stress of weather.
b. The part of a lower mast between the heel and the upper deck, or of the bowsprit between the stem and the knight-heads.
c. = ***house-line*****: see** house *n.* [1] **24.**
† **d.** ***housing-in*****: see** house *v.* [1] **8 (***obs.***).**
5. *Carpentry.* **(See quot.)**
6. *Mech.*
a. One of the plates or guards on the railway-carriage or truck, which form a lateral support for the axle-boxes.
b. The framing holding a journal-box. *spec.* **A massive metal frame or pillar that supports one end of a set of rolls in a rolling mill.**
c. The uprights supporting the cross-slide of a planer (Knight ***Dict.*** *Mech.* **1875).**
d. A structure that supports and encloses the bearings at the end of an axle or shaft; a journal-box. Hence more widely, a rigid case or cover that encloses and protects an axle or any other mechanism or piece of apparatus.
7. *attrib.* **and** *Comb.* **as** ***housing association, problem, project, question, reform, scheme, site, unit*****;**

Introduction(s)

The *pre*text for this book is the 25th annual meeting of the Society for the Study of Architecture in Canada, in Toronto. The book's *con*text is a bit different.

All academic conclaves require leavening. Normally, this involves at the very least a wine-and-cheese "meeter and greeter" to catch the early registrants, a licensed "banquet", and some time for conferees to wander off and sample the local trinket shops. But the SSAC moves its conferences to different locales each year so that the *place itself* becomes part of the conference. Thus, tours of the host city or town are meant to be integral to the conference rather than escapes. This is a good and important feature of the meetings, but there is the risk that by the end of the day the participants may be more exhausted than edified.

This little book was conceived to restore something of the "escape" function to the time outside of the conference theatre. Think of these essays and pictures as the guided tours that would have extended the conference by another week. And thus, this contribution *to* the conference becomes a legacy *from* the conference.

Plural voices

While conceived as a presentation about the conference's theme – "Fresh perspectives on housing" – this book is a really a selective and fragmented guide to locations and locales, presented in personal ways by a host of contributors. It has been a long time since a single voice could speak to the diversity and complexity of the centre of Toronto. This is a central city in the midst of a big city of almost two and a half million people, in the midst of a region of at least four million, and (according to some accounts) the anchor of a conurbation of more than eight million in two countries.

So we give you many takes, some *very* opinionated, on the housing of people in the midst of the city, viewed historically, geographically and architecturally. In the usual guided tour, one person holds a megaphone and tells stories to whomever is within earshot. In this tour, the megaphone is turned around, and sixty-five or so people are talking to one person – you, the reader.

To live in a city that works

The book is a presentation – more precisely a *re*-presentation – of a selection of buildings and sites and stories about living in the centre what was once described proudly – almost precisely 25 years ago – as a "city that works". You remember, don't you? The cover of *Time* magazine and all that?

Well, be assured that the city still does work. Like all cities, it works as well as it can.

Yet Toronto – a single entity embracing what had been six, and before that thirteen, and before *that*, dozens of towns, villages and townships – remains self-conscious and uncomfortable about how its image has withdrawn into the middle ranks of world cities since those euphoric 1970s. (The rest of Canada still doesn't like Toronto much; that at least hasn't changed.) The Toronto of 2000 faces crises of transportation capacity, social services and affordable housing. The well-managed halo of three decades ago has slipped. Indeed, the halo may have been yanked off altogether, whether by circumstance or more deliberate machination we dare not say.

Well (again), some of the brief commentaries in this book *do* say. On one side, there are new and fascinating places to live, renewed and revived neighbourhoods, and major efforts in research and planning for projects large and small. On another side, there are growing arguments for increased political autonomy for the city.

But on *another* side, there are people who have no fixed address.

Navigating east and west

In this book we look at east and west sides of "downtown." Until the expressway constructions of the 1950s, the map of built-up Toronto and its environs looked very clearly like an inverted **T**, hugging the shore of Lake Ontario, with Yonge Street its spine. Now that much of the rest is filled in, that central spine is no longer such a defining feature. So, to call this book *East/West* is fairly subtle (as well as a bit nostalgic). The subtlety is that street numbers in Toronto – even to the outer reaches of what people will still be calling Etobicoke and Scarborough for a very long time – start at the middle, at Yonge Street. In the middle of the centre of the city, the east side was developed and inhabited slightly before the west side, so *East/West* is chronologically valid. But on maps, the city reads left to right, thus *West/East*. (Attentive readers may find the title swapped around as often as not.)

Toronto is, after all, constructed on a mostly rectilinear grid of property subdivisions. One street is pretty much like another, so they say. Streets are usually straight, house lots are usually rectangles. What the preface calls cross-sections – what geographers and surveyors call transects – go north to south on both east and west. It's easy enough to navigate.

There are nonetheless several cautions for the visitor who wishes to walk, cycle, skate or drive these "routes." The "calming" of vehicular traffic is well established in predominantly residential areas.Many streets are one-way, and in the more assertive neighbourhoods they are almost no-way. Parking-meter rates and by-law enforcement have lately become more irritating, even during the brief time this book has been under construction. Pedestrians and other sidewalk users must contend with either frequent street crossings or long detours (cycling on sidewalks is for small children only, and there is still no consensus on where rollerbladers should be flying). So it's not so easy enough to navigate after all. Bring a map, and bring patience.

Making do

Throughout this book, and throughout Toronto's history, are case studies of *exigency* – mostly about making do under the pressure of existing conditions, with very rare (heroic?) attempts to overcome those conditions. Quite unlike Manhattan or other well-known examples of the parceling of land, the Toronto grid is in truth a set of nested grids. Just like those nested Russian dolls.

The original colonial survey comprised farm lots with roads almost a kilometre and half apart (100 surveyor's chains, or 6600 feet, to be more precise), and within each big square of this grid, 10 or 20 rectangular lots to be doled out or sold. Over two centuries, each of these big lots has been subdivided and subdivided and further subdivided, and none of the resulting properties nor even most of the roads were created or built on by government. The map of Toronto may be among the proudest expressions of 19th-century laissez-faire capitalism on the continent.

Roads that don't quite align, blocked or misdirected views, surprisingly wide boulevards, improbably narrow lanes – all these are part and parcel of the city's parts and parcels. Even University Avenue, the city's grand urban boulevard, is an accident of property ownership: there was a city street parallel to one of the University of Toronto's driveways, and the two were "assembled" only after many decades of inconvenience.

The oh-so-rare exceptions – the generous street allowances of parts of Spadina Avenue, and College and Queen streets – were private contributions to the urban landscape, from the beginning.

This leads directly to the theme of our little guide, the actual *housing* of people on these parts and parcels, these long straight streets and narrow lots. The neighbourhoods, blocks of houses, individual houses, and on down to their narrow yards and small rooms (*pace* Rosedale and Wychwood Park) – all follow from previous generations of subdivision. For that matter, so do their building forms and architectural styles, even the avowedly modernistic. All are adaptations. Even the most inventive – and there are many of these, when you look carefully – are founded on precedents, and they often look side-long at their neighbours.

This house, ⋯>, constructed on an eight-foot wide lot with an interior as *wide* inside as a normal residential door is *high*, may well be the most exigent example in the book, old or new. It was a neighbour and contemporary of one of those "Toronto residences" of 1934 shown at the very beginning of this book, though most of what you see now is a vertical addition of the 1970s. The architect/renovator went ahead with the conceit of expanding the narrow house even though he had acquired the 20-foot wide lot next door. This house is unusual, to be sure, but no less adventurous adaptations have come to exist throughout many older neighbourhoods, distracting us from what might have well been very homogeneous architecture and streetscape in their formative years. Such "marginal" cases give historic-district planners riotous nightmares – what does "integrity" or "authenticity" mean in such neighbourhoods? Perhaps "integrity" might actually be the opposite of "interest"?

(The house is not mapped, but it should not be so hard to find.)

Not making do

The exceptions to this general rule of exigency and adaptation are easy to spot: they comprise attempts to change the property lines. Land assembly – sometimes called blockbusting or urban renewal or even (more recently) urban *regeneration* – is the courageous attempt to reverse the historic course of progressively smaller increments of ownership and control. Land assembly is heroic because it is expensive – in terms of money and disruption – even on what might appear to be the poorest and cheapest lands.

The development of Regent Park, Moss Park, and Alexandra Park (public); or St James Town (private); or Harbourfront (confusing); or the West Don Lands/Ataratiri (failed) are all efforts at city rebuilding that erase and replace the texture of the city. Some of these erasures have proven successful over time, but it must be clear to even the blockbuster that the construction of new environments is not a science. Yet, the St Lawrence neighbourhood seems to be a successful – yes, *livable* – large-scale urban re-do. So, is it a coincidence that its overall pattern follows what had been the existing street plan, and that what became its central Crombie Park is evident in city maps of the late 19th century? Were its planners just *lucky*? Were the planners of St James Town or the central waterfront just *un*lucky?

In reading the streets and houses of central Toronto, what is evident on close inspection is the *malleability* of housing in a "traditional" neighbourhood over time, in terms of form, tenure, scale, even cost. In some respects this is a result of time; that is, over many decades there are many individual responses, and these are both cumulative and variable (with changing technology and fashion). Neo-traditional, "new urbanist" developments try to capture this image. Such projects are mostly missing from this edition (though, see Castle Hill, p 82, and the innards of the older blocks of the St Lawrence neighbourhood, p 68). This new "planning style" is quite conspicuous near the neighbourhood of the Candy Factory, p 153, or at the former old Woodbine/ Greenwood racetrack east of the Don River.

Such tracts tend to look very impersonal in their early years. Will they acquire the casual and varied character of the "typical" Toronto neighbourhoods, like the Annex or Sussex-Ulster or Cabbagetown in the very long term? Or does some factor of uniform construction, or density, or by-law strictures, or lack of non-residential land use preempt or impair such an evolution? Time will tell – particularly whether the new-urbanist style is able to help the establishment of "neighbourhood-ness" in this traditional sense. Portions of some neighbourhoods in this guide – Cabbagetown, for instance – did not reach the form we see today (more or less) for several decades. Others – many streets in Sussex-Ulster, for instance – were almost completely built up within a very few years in the 1880s.

Missing

Despite the variety of both subject matter and author, we have missed or passed over several more guidebooks' worth of lessons on urban housing in Toronto. For instance, we have not called attention to the quintessential high-density mixed-use pioneer – the Colonnade – as well as to the march of high-density residences north up Bay Street from Dundas, all in the image of the Colonnade, but far less impressive in quality than in quantity.

We haven't shown the quirkily modern pioneering live-work block at 477 Richmond West, though we have included several recent or impending "loft" projects, which are also less impressive in quality than in quantity.

We have passed over many examples of special-needs housing for seniors, or for the physically or psychically frail. We have not shown you what "modern" rooming-house conditions look like. We have not presented some of the physical solutions to the problems of poor or no housing for the economically or socially disadvantaged, like the Street City initiatives. At the other end, we haven't given you much of a look at those Georgianesque condos of the '90s, already quaintly dated, which may well become the middle-class housing of the upcoming '10s. We were overwhelmed by all the lofts. Sorry.

Coda

The planning and construction of housing – indeed, the planning and construction of whole neighbourhoods and entire cities – are at least as much (or more?) art, craft and politics as they are engineering, efficiency, and "common sense." Look closely at the places we propose, and at their neighbours as well.

It is both frustrating and refreshing to know that our predecessors (even in the 19th century) have grappled with problems of resources and techniques and politics no less difficult than the problems of our contemporaries – and that their tools and their solutions were just as reliable and no less likely to make good dwellings and good neighbourhoods. And we can still benefit from experience in other fields.

> Dr Gibbel of Lion Oil Electronics United addressed the stock holders, describing a new 'surprise' computer. After complex pre-programming by a team who does not know the computer's ultimate use, it is leased to someone who does not know the pre-programming routines.
>
> 'Ultimately,' he said, 'and without warning, the computer may do something either very stupid or very shrewd.'

So it may be for houses and neighbourhoods – and cities, too.

Mark Fram

Housing Toronto historically

History may inspire, caution, encourage, and constrain later generations. With respect to housing, the attitudes, institutional arrangements, and physical fabric we now confront have all evolved from earlier encounters with housing the City's residents. The following historical images illustrate the circumstances of real people in a variety of physical environments that reflect the entire social spectrum, from the unhoused and under-housed, to the modestly housed, and the distinctly over-housed.

The Homeless

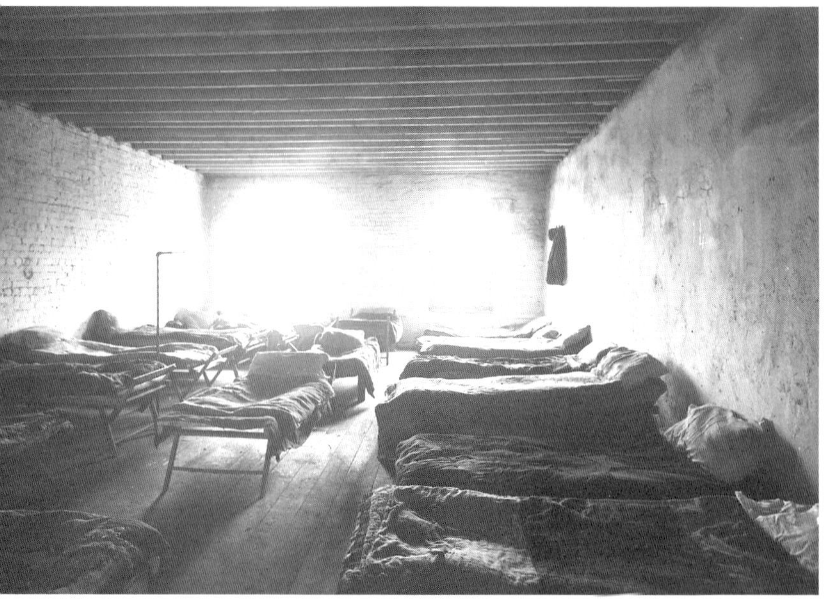

CTA: RG 8, Series 4, Sub-series 32, Item 6

Front Street Flop House, c. 1912

Labelled "10 Cent Lodging House on Front Street," this photograph was taken by the City's official photographer, Arthur Goss, in about 1912, to help illustrate the connection between poor housing and poor health. Dr Charles Hastings, the crusading Medical Officer of Health at the time, had some success in driving this point home and in promoting significant public health advances.

Boarders

CTA: RC 8, Series 4, Sub-series 32, Item 3

York Street Boarding House, c. 1912

Ever the consummate documentary photographer, Arthur Goss managed to persuade these Macedonian immigrants to come out of their tightly packed rooms to pose for his lens. Note the heating pipe snaking along the hallway ceiling, the battered wainscoting, and the decorative wallpaper adding a layer to paper-thin walls. The fact that the lodgings were "since improved with modern conveniences" was recorded on the back of the photograph. Just what these "modern conveniences" might have been is unknown.

Mrs Agnew's Boarding House, 244 Jarvis Street, c. 1886

No longer standing, 244 Jarvis was a common type of housing for its time: a brick, bay and gable, semi-detached house of modest proportions.

CTA: Fonds 80, File 3

Eating Watermelon, 244 Jarvis Street, c. 1886

CTA: Fonds 80, File 4

Mrs Catherine Agnew was a middle-class widow who, according to City records, bought and moved to 244 Jarvis in around 1875. Whether she was a teacher before her physician husband's death is not known, but she certainly taught at Winchester School in nearby Cabbagetown while residing at 244 Jarvis. In 1886 she was living with two named children (bank clerk Robert and law student John), as well as nine other unidentified boarders, some of whom may be enjoying the watermelon treat captured in this lighthearted photograph.

Public Housing

Spruce Court "Play Court," c. 1914

CTA: SC 18

In response to the inhumane housing found in such infamous parts of Toronto as "The Ward" (i.e., the area bounded by Queen, University, College and Yonge), the City's first government-sponsored housing was built at 74 to 86 Spruce Street in Cabbagetown. Constructed for working people, Spruce Court could hardly have been more different from the Ward and its ilk: solid, well-built, brick buildings; open, grassy common spaces; individual entrances; airy, well-ventilated, well-lit interiors;

modern gas stoves and electric fixtures; and careful detailing throughout. This photograph, looking south toward the old Toronto General Hospital on Gerrard Street, shows a group of children playing on the still-new greensward, supervised by adults from stoops, porches, and verandahs looking out over the common area. At this time, Spruce Court contained 32 cottage flats and 6 six-room houses.

CTA: SC 18, Item 26

Spruce Court, Living Area, c. 1914

"Sanitary conveniences, adequate bedroom accommodation and domestic privacy are the primary requirements for proper housing," proclaimed the Toronto Housing Company's 1915 publication, *Cottage Flats*, where this photograph by William James first appeared. At that time, "cottage flats" ranged in size from small, one-bedroom flats to four-bedroom, two-storey units, and monthly rents ran from $14.50 to $29.00, depending on size and location. This photograph shows the living room in one of the larger cottage flats at Spruce Court, furnished as a model suite. The layout and the furnishings reflect the aspirations of the Housing Company to house lower-income wage earners well.

Unfortunately, this early housing experiment was not a financial success. At a time when the average wage was still considerably under $15.00 a week the rents charged by the Toronto Housing Company proved too onerous for the target clientele. Just twenty years later, in the middle of the Great Depression, the Toronto Housing Company ceased operations. Fortunately, the physical fabric of the projects was neither razed nor renovated into oblivion. Spruce Court now operates as a successful City housing co-op.

The Middle Classes

Tea Time at 39 Huron Street, c. 1909

CTA: SC 244, Item 3554

William James, the stern patriarch at the centre of this large family grouping, emigrated from England in 1906. By 1909 he was an established professional photographer and living in the first of his four Toronto homes, 39 Huron Street, a tiny, worker's cottage located south of Grange Avenue. The James family lived in crowded conditions: here we see eleven people (and one cat) gathered around a foldout table in the corner of the kitchen/dining/living room. Still, all the proprieties are being observed: a proper English tea is set on a white tablecloth; solid furniture rings the room; knick-knacks are neatly displayed along the mantelpiece; leather-bound books are piled atop the highboy; and the loud waterlily wallpaper is decorated with an abundance of prints, photographs, and calendars. The building still stands and is still used for housing.

Parlour Portrait, 697 Spadina Avenue, 1893

CTA: SC 128, Series 1, Item 1234

This parlour portrait is devoid of people, but full of information about both the house and its occupant. Although bigger than the James cottage and located in a more affluent neighbourhood, this Romanesque

Revival residence is still modestly proportioned. Through the string curtains one can see that the main staircase is a switchback, rather than a straight bank, and that the parlour and dining rooms open off either side of a small entrance hall, rather than directly into one another as in many semi-detached houses. The kitchen is tucked discreetly out of sight in the back.

As for the parlour itself, it speaks volumes about the activities, tastes, and even aspirations of its occupant, or possibly its real estate agent, since the first occupant, Stacey Lake, a Gutta-Percha Co. cashier had only just moved in about the time the photograph was taken. Everything is laid out and carefully displayed for the camera: the ubiquitous parlour piano and wildly popular stereoscopic viewer; a large, five-string banjo, and issues of *Scribner's*. The lightly coloured walls lessen any claustrophobic feeling, and the stylish gasolier hangs ready to cast light whenever it's needed. This residence still stands on the east side of Spadina Avenue, just south of Washington Avenue.

Upper Classes

Blake Residence, 467–469 (formerly 397–399) Jarvis Street, c. 1880

CTA: SC 146, Item 2

The Hon. Edward Blake, who was the first Liberal Premier of Ontario (1870–71), purchased 397–399 Jarvis Street ("Humewood") in 1879. Standing on the east side of tree-lined and ultra-fashionable Jarvis Street, Blake's picturesque, Second Empire-cum-Italianate residence was actually a pair of semi-detached dwellings that he adapted according to the needs of his family. Walls were knocked out and filled in and front doors were shifted, rejigging both interiors and exteriors as deemed necessary. Blake also purchased the house immediately to the south for other members of his family. Now these large residences are occupied by businesses; accountants and small offices are in Humewood, and the Red Lion pub is in the house to the south.

Blake's Study, 467–469 Jarvis Street, c. 1880

Most photographs of Victorian interiors show highly decorated and overstuffed spaces but this upstairs room looks comfortably lived in. Blake's study gives the impression that the great man has just stepped out for a moment and will return shortly to work on the legal papers strewn across his desk. The sun-drenched window faces south, overlooking what is now the Red Lion and an empty chair appears to await the arrival of a statesman or favour-seeker to consult with the Liberal Leader.

CTA: SC 146, *Item 11*

CTA: SC 155, *Item 1*

Cumberland House, "Pendarvis," 33 St George Street, late 1880s

Frederick W. Cumberland built "Pendarvis" between 1857 and 1860, while creating his near-by masterpiece, University College. "Pendarvis" was the first building on St George Street. It served as Cumberland's home until his death in 1881 and is now generally known as Cumberland House.

The mansion was redesigned in 1883 by his partner, William Storm, and remained in residential use until the University of Toronto purchased the property in 1923. Since then, this handsome, mid-Victorian mansion has been renovated and put to new use several times. It now functions as a lively International Students Centre.

The most obvious changes since Cumberland's day concern the siting of the building, which is now hemmed in on three sides and nearly invisible from any point except St George Street. When it was first built, Cumberland House was entered from the east (now a parking lot); it looked southward across fields toward the distant city (a view now blocked by the Wallberg Engineering Building); and it turned its back on the new St George Street. At the time of this photograph, the main entrance appears to be in the centre of the south facade and a delicate verandah extends along the building's entire length. The house was still visually prominent and approachable from several directions. Today the main entrance is on the west side, once the back of the house.

Hall, Cumberland House, 33 St George Street, c. 1890

CTA: SC 155, Item 4

No photographs survive of the original Cumberland interior; however, there are some taken by J. Bruce when the house was about 30 years old. In contrast, the latest decoration of the International Students Centre displays elements of neoclassical restraint. For example, the walls of a meeting room running along the south facade are painted creamy yellow, and the ceiling, decorative plaster, mouldings, pillars, and elegantly wrought radiator covers are all painted white – a far cry from the late Victorian opulence shown in this photograph. Here we see a large, but dark and claustrophobic Victorian hallway. Statues, urned plants, and heavy mirrors crowd the passage. Every wall and ceiling is hung with patterned paper. Even an internal doorway is draped with fashionably heavy swags.

Sally Gibson

EAST

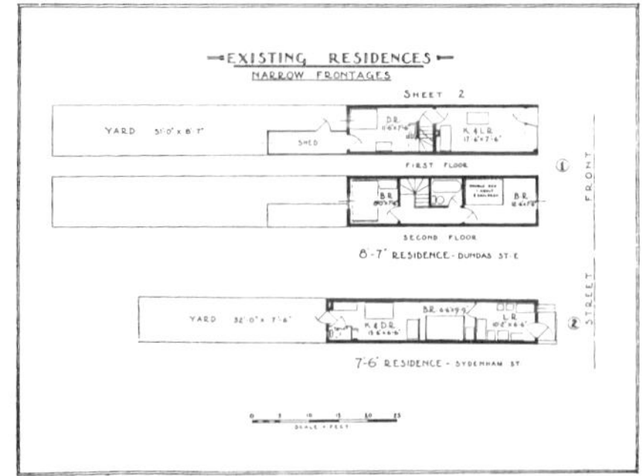
EXISTING RESIDENCES
NARROW FRONTAGES
SHEET 2
YARD 31'-0" x 8'-7"
SHED
D.R.
K. & L.R.
FIRST FLOOR
B.R.
B.R.
SECOND FLOOR
8'-7" RESIDENCE - DUNDAS ST. E
1
FRONT STREET
YARD 32'-0" x 7'-6"
K. & D.R.
B.R.
L.R.
7'-6" RESIDENCE - SYDENHAM ST
2
0 5 10 15 20 25
SCALE - FEET

18
Rosedale

25
St James Town

30
Metcalfe Street

38
Sherbourne/Dundas

44
Regent Park

48
Trefann Court

62
Corktown

64
West Don Lands

68
St Lawrence

Rosedale

Rosedale, which has loomed large on Toronto's mythic landscape for nearly 150 years as home to many wealthy and well-known people, occupies an impressive 620 acres near the centre of the city, and contains over 2,200 dwellings. Bounded on the west by Yonge Street, the CPR rail line on the north, the Don Valley on the east, and by Bloor Street and the Rosedale Valley on the south, Rosedale is split by the Silver Creek or Park Drive ravine into a northern and southern part.

South Rosedale had its origins in the 1854 subdivision of part of Sheriff William B. Jarvis's estate on the edge of Yorkville Village to create 61 villa lots within a pattern of winding streets that responded to the adjacent ravines rather than a grid. This plan, highly unusual for the time, originated with Jarvis's surveyor, J. Stoughton Dennis. Later, between 1877 and 1896, it was extended over three other adjacent but smaller estates. From these four estates only one great-house survives (5 Drumsnab Road, 1835; altered 1856). Villas dating from the first wave of building in Jarvis's subdivision include 23 Rosedale Road (1857; altered 1911), 124 Park Road (1857), "Glen Hurst" (1866) on the grounds of Branksome Hall Girls School, 27 Rosedale Road (1871), and 3 Meredith Crescent (1876). When Rosedale, as part of Yorkville, was annexed to the City in 1883, a few new houses stood on Elm Avenue and South Drive, but not many elsewhere. Along the ravine that separated South Rosedale from the near-empty lands to the north, three large mansions had taken up sentry-like positions flanking Glen Road and Elm Avenue. Today only shady grounds and handsome gates (1903) of "Craigleigh" survive to recall that trio.

Between 1884 and 1904 development was slow and scattered. No more than three dozen houses were constructed then, of which 2 Dale Avenue (1887), 128 Park Road (1893), and a terrace at 141–147 Roxborough Street East (c.1890) are good examples. The great boom in South Rosedale, which accounts for much of its present architectural character, didn't occur until 1904–14, when a couple of hundred dwellings were built. Streets like Chestnut Park were filled up then with big, brick houses, many in an Arts-and-Crafts style, by architects like Burke, Horwood and White (No. 1, 1915), A.E. Boultbee (No. 20, 1905–06), S.H. Townsend (No. 24, 1905), and E.J. Lennox (No. 48, 1903). What was for many years the area's only apartment house was erected at 75 Crescent Road in 1912. At that time, too, more boulevard trees were planted that now contribute so much to Rosedale's garden-suburb appearance.

Meanwhile, North Rosedale had begun to grow. An iron bridge erected in the early 1880s to carry Glen Road over to the Silver Creek ravine did little to stimulate building there until shortly before the First World War. Development was helped by the decision in 1911 to erect "Chorley Park," the palatial and now-demolished official residence of Ontario's Lieutenant Governor, on a 14-acre site overlooking the Don Valley between Roxborough Drive and Summerhill Avenue. About that time, a few large houses were erected on Binscarth Road, Highland Avenue and Beaumont Road, and several smaller ones on streets north of Summerhill Avenue to the CPR right-of-way. But there was no building on the lands between, where the Lacrosse Grounds (now Rosedale Park) and St Andrew's College were located, until after the school removed to Aurora in the 1920s.

Thanks to redevelopment of the George Estate as Old George Place, and other smaller projects, North Rosedale boasts some excellent examples of modern-period architecture by Ron Thom (4 Old George, 1971), John B. Parkin (3 Old George, 1959), and Barton Myers (51 Roxborough Drive, 1972).

E/R

In some ways, Rosedale led a charmed life in the post-war period, only to suffer more recently with the architectural excesses of Big Money. When Mount Pleasant Road was extended through the area to link up with Jarvis Street, only two houses and a worn-out school were demolished. A spate of apartment building in the 1950s, taking advantage of large ravine lots, was brought under control before much damage was done to the area's character. And the Crosstown Expressway, linking the Don Valley to Davenport Road, died on the drawing board. Whether Rosedale will emerge with its integrity intact from the current wave of redevelopment, characterized by a taste for neo-Georgian country houses squeezed onto city lots, remains to be seen.

Stephen A. Otto

Ancroft Place

Sherbourne Street at Ancroft Place
Architects, Shepard and Calvin
Completed 1927

Toronto has few 75-year old housing projects with the urbanity of Ancroft Place. Conceived in 1926 as a way to redevelop a large, old estate with a minimal frontage on the east side of North Sherbourne Street above Bloor, it consists of 21 houses grouped into three blocks. Five dwellings face the main street (Sherbourne), while 16 others have numbers on Ancroft Place, a side street close to the south edge of the property. The units vary in size

E/R

from six to nine rooms. Behind the buildings a lane leads to the service entrances and heated, private garages that are attached to each unit. The architects for the complex, Shepard and Calvin of Toronto, succeeded brilliantly in creating an informal pattern of clustered residences in the English cottage style, where none looks directly into the windows of a near neighbour. While appearing as separate dwellings, the houses are in fact in a single ownership and share a common heating plant.

Stephen A. Otto

Selby Hotel

592 Sherbourne Street
Architect, David Roberts Jr
Completed c.1883

The Selby Hotel was originally home to the Gooderham family, distillers extraordinaire. A prime example of Victorian architecture, it features wrap-around verandahs, gables, bargeboards, dormer windows, and exceptional stone and brickwork. The architect, David Roberts Jr, was the same architect who designed the mansion at 504 Jarvis Street in 1891, the Flatiron building at Front and Wellington (the original head office for Gooderham and Worts), and the York Club on the northeast corner of St George and Bloor.

Sherbourne Street, like the parallel Church and Jarvis streets, was at one time the most fashionable street in Toronto. But the area experienced a decline as other residential neighbourhoods, such as the Annex and Rosedale, grew. Upkeep was especially onerous on these large properties. Only in the 1980s did signs of revitalization begin to be seen, as the once elegant mansions became commercial spaces incorporating offices, restaurants, rooming houses, and convenience stores. Despite the fact that it presently serves as a budget hotel, the Selby is fortunate to have largely maintained the scale of its original spaces, although the visitor has to work to imagine it as a luxury residence.

EJR

The Gooderhams lived in the house from 1894 to 1910, after which their home became the early incarnation of Branksome Hall, a private girls' school, from 1910 to 1913. Thereafter, the building became known as the Selby Hotel, so named as it is situated on the corner of Sherbourne and Selby streets.

A classic H-plan configuration was added to the western portion of the property to increase capacity. The Selby served as an early budget hotel and was seconded as a home for World War II officers. At one time Ernest Hemingway stayed at the Selby with his wife Hadley, who occupied an adjoining suite next to her husband. By the 1970s the Selby had become seedy as prostitutes were using its main floor rooms for clients.

The Selby continued to slip into decline until Rick Stenhouse purchased the hotel in 1984 and went about the process of restoring and preserving the property to its original character. The suites in the original home were restored to their 15-foot ceiling heights, and mouldings, doorways, and fireplace mantels were all restored. The cast-iron fence bordering the property is not original but it is similar to the fence that protected the original Gooderham home. The slate roof was restored in the mid-1980s. Today, the well-known gay bar "Boots" occupies the basement of the hotel and the Selby attracts both locals and tourists who are drawn to the offerings of the neighbouring gay community.

Ian Chodikoff

Peggy and Andrew Brewin Housing Co-operative

79 Charles Street East
Quadrangle Architects Limited
Completed 1994

This project was designed within the restrictive building envelope of a previously approved condominium tower for the site. The building is a co-operative housing development, which includes housing for persons with AIDS. The project consists of a 66 suite, 16-storey, stepped point tower on Church Street and a 20 unit, four-storey, walk-up apartment along the rear lane, with a landscaped courtyard between the two structures. The tower has a strong cornice line at the fourth floor that ties together the two buildings, and the rear building reinforces a City policy aimed at improving public safety by putting life onto the back lanes.

Leslie M. Klein

• *Quadrangle Architects*

Homewood

North side of Wellesley Street, between Sherbourne and Wellesley Place
Architect, Henry Bowyer Lane
Completed 1847; demolished 1964

The most obvious sign that remains of Homewood is a bend in the road. To most people travelling on Wellesley Street East, this gentle curve is an odd and inexplicable diversion in Toronto's otherwise straightforward grid system. It isn't a surveyor's error (an incorrect meeting of two grids), but an indication of the pattern of early settlement in the area. The land between Queen and Bloor streets had been originally laid out as large estate lots, called park lots, which were gradually subdivided to accommodate urban development. The Allen family owned this park lot, and it was the younger Allen, George, who built his villa Homewood on this site. Homewood, which had been built in 1847 to the designs of the noted architect Henry Bowyer Lane, remained standing until 1964. In 1900, it was the first house in Toronto to have electricity, and when its grounds had been reduced to 4 1/2 acres, it was converted into the first Wellesley Hospital, which was formally opened by Sir Wilfrid Laurier in 1911.

The gentle curve in the road marked the gracious front yard of the estate. Homewood Avenue created a central, axial corridor from the house down to Allen Gardens, the horticultural grounds donated by the Allen family to the city as the central focus of their park lot development. This corridor continued further south as Pembroke Street, leading to Moss Park, the estate of George Allen's father. This plan of linked gardens and estate lots with landscaped lawns, orchards, greenhouses and tennis courts gradually became the framework for the existing city fabric. It underlies this part of Toronto, like an almost forgotten memory, a dream of a pastoral city.

Michael McClelland

St James Town

Bounded by Sherbourne, Howard, Parliament and Wellesley streets
Architects, George Jarosz and James Murray
Completed 1965-68

With a population of about 15,000 people on an area just over one fifth of a square kilometre, St James Town is probably the most densely populated piece of real estate in the whole of Canada. Not only are there a lot of people living in St James Town but they are a very mixed group of people and many are new immigrants (St Jamestown represents one of the major concentrations of new immigrants in the city). Rose Avenue Public School, the local school, boasts that it has the most diverse group of students anywhere in the Greater Toronto area. There are children here from dozens of different countries.

One of the unique aspects of St James Town in the Toronto context is that its 18 high-rise buildings are owned by a mix of public and private landlords. Four of the larger buildings are owned by Metro Toronto Housing Authority, and the other 14 buildings are privately owned. Many of the privately owned buildings were seen as "up market" addresses when they were first occupied in the 1960s. A number of them still boast swimming pools in their basements, although most have not been used for years.

E/R

St James Town is still seen by many of its residents as a good place to live, since it is centrally located, rents are affordable, and, although there is a paucity of recreational facilities within St James Town itself, the diversions of downtown are nearby. When the high-rise towers of St James Town were built in the early 1960s, they displaced the kind of housing that takes up much of the current South St. James Town, i.e. that area south of Wellesley and down to Carlton Street. Obviously, the displacement of detached houses, semi-detached houses, and row houses has had a marked effect on the urban landscape, and there was considerable protest against the change when it happened. Even today, many feel the scale of development is inappropriate. However, against this one must weigh the fact that many more people are now within easy access of the downtown area. And any discussion with the residents of St James Town will reveal that there are benefits to the high-rise life, not least of which are the views.

Jim Ward

St James Town reconsidered

1975, St James Town, a summer day. A group of swingers in bell bottoms, carrying beer and pizza, pop into one of the 30-storey-high or so apartments in the area to join a weekend swing party. These apartment buildings were occupied by young singles or couples who were proud to say, "Yeah, we live in St James Town!"

1999, St James Town, summer day. A group of young mothers, whose first language is not English, push their baby strollers to the corner of apartment 135, waiting for the school bus that will bring their children to a local Catholic school. They were talking about the break-in at Mrs Tan's place, an apartment unit that was home to two adults and six children. When asked if they would bring their kids to one of the few playgrounds in the neighbourhood, most said, "Probably not, unless the place has other activities for the teenagers so that they won't take over the playground."

These are the concerns that inspired a group of 20 University of Toronto landscape architecture and architecture students to seek ways to help improve this neighbourhood. On February 9, 2000, the team led a group of 30 (residents, planners, and public officials), on a walk around the neighbourhood, recording their observations about the site and discussing ways to improve it. The following are some of the comments that were made:

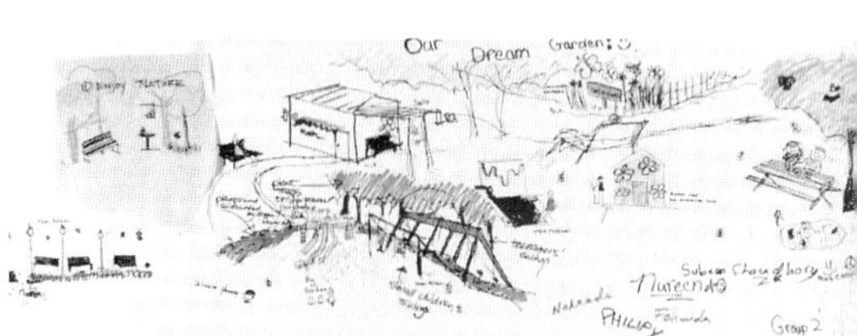

"More lighting [is] needed and signs for the school kids on the road; people do not bring dogs [on a] leash and they poop all over the place."

"When crossing the street at Wellesley and Ontario I'm asked questions [about drugs]."

Christina Tang

"[The service lane] looks very dangerous, puzzling, you don't know where [it's leading] to."

"I think we should turn down all the fences and change the service lane into a park and put the community garden somewhere where everyone can use it."

"We should apply for grants to hire the local teenagers to clean up litter and guard the park during the summer ... This will use their energy positively, and educate them to care about their community."

Yvonne Sze-Man Yuen

South St James Town

Bounded by Sherbourne, Wellesley, Parliament and Gerrard streets

South St James Town retains much of the physical character that was once present in St James Town itself. The housing is mostly two-storey and of 19th-century vintage. Many houses have been rooming houses for several decades, one of the major landlords being the City of Toronto Department of Housing; which owns several clusters of rooming houses along Wellesley Street, Parliament Street, Prospect Street, and Carlton Street. These rooming houses accommodate several hundred low-income earners close to the downtown area. There are also several denser, affordable-housing developments in South St James Town; some are City-owned and the Hugh Garner Housing Co-operative forms an important presence in the northwestern corner of the area. In at least one case, a City-owned building has been made into a co-operative housing project.

E/R

Although only Parliament Street separates South St James Town from Cabbagetown to the east, South St James Town is markedly different in socioeconomic terms. In 1996, for example, median family incomes were less than half those in Cabbagetown. As with St James Town itself, this part of Toronto provides fairly low-cost housing in an area close to the city centre.

Jim Ward

Residential high-rise buildings

Behold the lowly, but tall, residential high-rise building. A thousand buildings like this cover the city – a promotional booklet from the period proclaims that Toronto built over 200 buildings of this size in 1968 alone. These buildings are supreme examples of modern planning rationalism. To paraphrase Sprint Canada, they get "the most for the least." The most people living on a plot of land with the least possible amount of material, that is. The material restrictions dictate a certain severity and optimum deployment, which results in a limited number of stylistic options – all of which have been played out around Toronto over time. One of the standard styles is "Rectilinear Classic":

Stage 1 –

- Level the site of all that came before;
- Excavate and lay in as much parking as possible;
- Relay a thin layer of grass, and include soon-to-be-smelly stairwells poking out from below;
- Fence the site (optional);
- In the midst of this, situate a tower, slab walls aligning with the parking layout below;
- Extrude stories as required and/or allowed (whichever comes first); with absolute uniformity from bottom to top.

Stage 2 – consider options: How to deal with the balconies? What material for infilling the concrete frame? How should the building hit the ground? (Toronto architects are notoriously good at avoiding the relationship of building to ground.) How to deal with all the exposed concrete?

Rectilinear Classic makes the obvious choices:

- Symmetry, symmetry, symmetry;
- Solid balcony guards (solid appearance = clean appearance);
- Pinch the corners to take the bulk out (i.e., don't put balconies at the end);
- Reveal the concrete structure by infilling the walls within the structural frame (why pay for covering the structure, and it adds visual interest);

It was within these parameters that the high-rise housing game of the 1950s, '60s, and '70s was played. After 1980, the rules changed again. When is the last time you saw exposed concrete? Looking back from the 21st century, we now know these buildings, with all their faults and their energetic ideals, are now historic in their own right.

Ian Panabaker

Paul Kane House

56 Wellesley Street East
Built 1854–56
Addition: Architect, Paul Reuber
Completed 1986

Designed in 1983 and completed in 1986, this Lilliputian 18-unit apartment block was the third and final phase for the community-based Church-Isabella Residents' Co-operative Inc. It was the first project of substance by its architect.

Paul Kane, famous for his 19th century oil paintings depicting North America's native people, built a Georgian home on the site in 1856. Later, to please a more Victorian eye, a side bay window, a two-storey porch, and a peaked roofline were added. In the 1930s, the house was completely obscured by the construction of an addition to house a church for the deaf. A developer purchased the property in the mid-1970s, intending to build a high-rise structure. The church was demolished and, suddenly, the long-forgotten Kane house was again exposed to public view.

Robert Mikel/THB

Community pressure prevented the ultimate destruction of the Kane house, and the City eventually purchased the property, transferring the developer's density rights to an adjacent site. In 1983, after the house had suffered several years of vacant neglect, the Church Isabella Co-operative successfully answered a City proposal call with a scheme to lease the rear of the site for residential use and landscape the front as a public park.

The front wing of the house has been renovated to contain multi-level three-bedroom units. The fire-damaged rear wing has been demolished to make way for a four-storey addition containing eight lower units with tiny rear gardens and eight upper units with roof terraces. Corridors on the first and third levels are singly loaded and have operable windows facing the park. There is a kitchen window surveying the corridor next to the front door of every unit. The front facade expresses the collective nature of the co-operative, while the rear expresses the individuality of each unit.

The rehabilitation of the Paul Kane house addressed three goals: it found a reuse for a historic building, provided affordable family housing close to a subway station, and created a public park in an area chronically short of green space.

Paul Reuber

8 Wellesley Street East

Architect, Dermot J Sweeny
Conversion completed 1996

The recession of the early 1990s left many vacant or under-leased commercial buildings in the downtown. Office conversions to residential use became a City-supported developer's rally – in mid-1995, according to the (old) City of Toronto's Planning Department, there were at least 20 applications for conversions. Eight Wellesley Street East was one of the first conversions of a 1960s era, Miesian, curtain-walled, waffle-slabbed office building to an 81 unit condominium, and among one of the

EJR

most successful. Directly across from 8 Wellesley Street, at 555 Yonge, there is a similarly proportioned building – a contemporary conversion – that is visually much less satisfying. One wonders whether, if the same developer had bought both, there might have been a twinned gateway to Wellesley Street East and the neighbourhoods beyond.

David Winterton

Metcalfe Street: the do-it-yourself approach to heritage conservation districts in Cabbagetown

In a community such as Cabbagetown, it's not surprising that the impetus for a Heritage Conservation District (HCD) came from the "grassroots." After all, Cabbagetown has a well-established preservation association, a deep sense of community pride, and a unique architectural heritage often touted as "North America's largest collection of Victorian homes." Yet, in Toronto, HCDs are unusual – and a grassroots approach to creating them is even more unusual. So far, only four HCDs have been approved in the city, and all were "top-down initiatives" proposed and implemented by Heritage Toronto back in the "glory days" when staff and resources were plentiful.

In 1995, Peggy Kurtin, then President of the Cabbagetown Preservation Association, began working with a dedicated group of individuals to put together the initial study to establish the HCD. She says, "We knew it would be an uphill battle because it hadn't been done this way before." Fortunately, the volunteer group had enormous tenacity. The study had been limited to Metcalfe Street, one of the most charming and best-preserved streets in Cabbagetown. Metcalfe Street exudes a sense of cohesiveness, probably because it features a uniform style of wrought iron fencing due to a communal agreement in the past. In both a cultural sense and an architectural sense, Metcalfe Street was an excellent candidate for a HCD.

Although the volunteer group was delighted by the warm endorsement in official circles, a proposal was made to expand boundaries that also expanded the workload. Kurtin says, "Most people have no idea how time-consuming this process is – every single house in the area must be researched at the City Archives and cross-checked at the Registry Office, and then catalogued with a formal written description and photograph." The project kept a corps of twelve volunteers working for months.

In addition to historical research, the process of gaining official status as an HCD brings many other challenges. One of the most delicate is what is sometimes called "managing the public process" – making sure that resident homeowners are well-informed and invited to provide their comments and express their concerns. The ideal goal is to gain unanimous support from homeowners in the area.

The delicate, and somewhat political issue, is homeowners' rights. When an area becomes an HCD, homeowners must seek special approvals to make architectural alterations which must generally conform to the historical guidelines that have been established for the district. In the Metcalfe Street proposal, it was decided that guidelines would be limited to the exterior facades of the homes and the streetscape. In other words, homeowners would be free to make whatever interior renovations they wished, assuming they complied with standard building regulations. As part of the process, a

Steering Committee was formed to include community representatives from each of the five streets in the expanded area, and representatives from key official sectors. Public meetings were held and went surprisingly smoothly. In fact, Kurtin says the only complaint they heard was from one homeowner outside of the proposed district, wondering why his home hadn't been included!

Clearly, in the case of the proposed Metcalfe Street HCD, homeowners recognize there are many benefits, some of which are financial. Studies suggest that real estate values generally go up. In Toronto, municipal heritage staff offer grants for historic renovation and supply some cost-free advice and expertise.

Peggy Kurtin

Despite the enthusiasm for the Metcalfe Street HCD and unanimous support from all homeowners in the area, it is not yet an official reality. Assuming the proposal gets the right blessings along the way, it must still go to the Ontario Municipal Board for final approval.

As Kurtin says, "It's an onerous process – not something you do lightly." When asked why she has put so much time and energy into it, she says, "I chose to live here because I liked the character of the area. That's what attracted me to the area. I don't want that to change."

Kathy Farrell

Spruce Court

(previously known as
Sumach Street Terraces)
74–86 Spruce Street
Architects, Eden Smith and Sons
Completed 1913

Spruce Court represents the first deliberate attempt to create low-income public housing in Canada. It was the product of an early public and private partnership initiative, which identified the need for low-income housing, arranged for financing, and administered the project until 1980. Spruce Court is also notable because it exhibited the design attributes, architectural style, and high quality of construction usually associated with single-family housing.

Wins Bridgman

The complex is located at the corner of Spruce and Sumach streets, near Dundas and Parliament. Spruce Court comprises 78 various-sized flats in two- and three-storey buildings. All the flats have a front door and porch to the outside. They face either Spruce Street or Sumach Street, or face one of two courtyards open on one side to the street. The first half of the project formed the Sumach Street courtyard in 1913. The Spruce Street courtyard followed in 1926. Riverdale Courts, constructed in 1913, was a larger but similar project on Bain Avenue, also designed by the architectural firm Eden Smith and Sons.

The funding and planning for Spruce Court was arranged by the Toronto Housing Company (an earlier version of Cityhome) in 1913, in response to a need for rental housing closer to the industrial heart of the city. The effort involved a partnership between the business community and the Canadian Manufacturer's Association, with the Toronto Board of Trade, the Guild of Civic Art, the Local Council of Women (who were especially concerned about the welfare of working single mothers), and the Toronto City Council.

Eden Smith (1858–1949) was a sought-after Toronto architect. Most of the 2,500 single-family houses he designed and constructed were built in the English Cottage Style. Concurrent with Spruce Court, for instance, Smith was involved in the planning of the site, and the architect for six notable houses, in what is known today as Wychwood Park.

The English Cottage Style used at Spruce Court evoked in the popular mind the image of a simple and domestic, country life. This romantic association took the form of steep shingled roofs, broad eaves, tall chimneys, brick and stucco walls, half- timbered gables, large roofed verandas, arched brick porches with stone steps, heavy wood front doors, and, within the site's restrictions, verdant courtyards and gardens. Equally important to the character of the buildings are the quality of materials, for example, the

Wins Bridgman

solid wood doors, the heavy brass hardware, the stone thresholds in the porches and the play of voids and solids. The latter is created by the shadow in the porch archways, the roofed verandah, the shadowed eave on otherwise simple building volumes, and the large and divided windows with deep reveals. The interiors also contain enduring materials, such as hardwood floors, decorative wood mouldings, doglegged hardwood stairs, and plaster walls. Interior and exterior details and rich materials were used sparingly but to great purpose.

The Spruce Court buildings were an immediate success with the new residents, and they continue to provide a rare level of domestic pleasure in public housing.

Wins Bridgman

Three Streets Housing Co-op

77 Winchester Street,
39 and 43 Metcalfe Street

This 39-unit apartment building became a non-profit housing co-operative in 1981, after a high-spirited tenant battle with the owner and the Canada Mortgage and Housing Corporation (CMHC). Built in 1911, the former Hampton Mansions was a gracious, middle-class apartment house with fine chestnut woodwork, tiled gas fireplaces, large windows, good cross ventilation, and dumb-waiters connecting each unit to delivery areas in the basement. Hampton Mansions and a number of neighbouring properties on Winchester, Metcalfe, and Sackville streets had been assembled during the late 1960s and early 1970s by a developer hoping to follow the St James Town trend. The rise of neighbourhood activism saved the Don Vale neighbourhood from "renewal" and the properties were held as modestly priced rental accommodation into the 1980s. When a small consortium of business people bought the properties in 1980, the tenants realized that their homes were heading for luxury conversion. A tenants' association was formed and a battle plan developed to save as much affordable rental accommodation as possible. When the dust settled nearly two years later, the tenants had achieved a significant victory – Hampton Mansions had been purchased through the CMHC co-operative housing program. The name "Three Streets" was chosen in recognition of the original scope of the tenants' ambitious campaign.

E/R

Hampton Mansions had been constructed in stages as three separate, three-storey buildings. When it became Three Streets, extensive renovation was carried out, with attention paid to restoring and preserving the historic elements of the building. The exterior yellow brick was carefully cleaned; the original woodwork in the hallways and staircases was painstakingly refinished; unit rewiring was accomplished with minimal damage to existing plaster; replacement iron work was designed to retain the Hampton Mansions decorative elements. One of the most exciting improvements has been the creation of a "secret garden" in the large interior courtyard. The members also added an ample roof deck that has stunning views of Toronto in all directions.

Today, Three Streets houses individuals and families at both market and subsidized rents. Through an agreement with the AIDS Committee of Toronto, one subsidized unit is reserved for a member who is HIV positive.

Cynthia Wilkey

City Park and Village Green

City Park Apartments: 484 Church Street
Architect, Peter Caspari
Completed 1954

Village Green: 40 and 50 Alexander Street,
55 Maitland Street
Architect, John Daniels
Completed 1967

Two years after the 1952 completion of Le Corbusier's *Unité d'habitation* in Marseilles, High Modern residential architecture arrived in Toronto in the form of the City Park Apartments at 484 Church Street, just north of Maple Leaf Gardens. This, the first post-war apartment complex in Toronto, replaced the Victorian houses on its site with three 14-storey slab blocks containing 774 rental apartments.

"*Espace, soleil, verdure*" is provided for all residents, thanks to the east-west orientation of apartment units and generous spacing between buildings. Apartments are large, balconies private, and waiting lists are long. Following conversion to a non-profit housing co-operative in the early 1990s, City Park underwent renovations, including the application of vertical metal cladding on the north and south end walls of each block.

Thirteen years after City Park was built, the 700-rental-unit Village Green project was constructed across the street to the north. The three buildings – two 19-storey slabs (with two-storey penthouse units on top) and a cylindrical 28-storey point tower – are clad in white and blue glazed brick. Village Green provides lots of *espace* and *verdure*, but north-facing units lose out on *soleil*.

EJR

Both City Park and Village Green are examples of high-rise, high-density site planning at its best. An interesting comparison can be made with the two most recent developments in the immediate vicinity. The Alexus, at the northwest corner of Church and Alexander, is a condominium apartment building built to the property line that steps up from one to eleven storeys. It has a false front on Church Street (the cornice line of which unfortunately doesn't match its older neighbour to the north). At 42–52 Maitland Street an infill development of 20 condominium townhouse units is packed so tightly on its site that rear windows are almost non-existent due to building-code restrictions. The high density of these and many other residential buildings in the area supports the busy Church Street commercial district, a neighbourhood shopping area, and GHQ for Greater Toronto's gay community.

Douglas Young

Merchandise Building

108 Mutual Street
Current architect, Paul Northgrave
Completed 1910
Altered 1916, 1930, 2000

In 1910 Robert Simpson Co Ltd constructed on the southwest corner of this site a five-storey, reinforced concrete and steel structure, which was used for wagon storage and as a harness shop. In 1916 an 11-storey mail-order building and warehouse was added at the south end of Mutual Street, designed by Max Dunning and Burke, Horwood and White. The final construction phase of 1930 took place at the north and west sides of the city block. The design of the west facade is a fine example of multi-storey warehouse design. Over 12,000 people worked in this huge industrial building at one time.

In 1977 construction began on a new; mixed-use development (residential, retail, and commercial), after interior elements of the original building had been removed, except for concrete columns and floors. The Merchandise Building now consists of 504 loft residential units, 529 above-grade car parking spaces, 246 bicycle spaces, 4 loading bays, a 30,000-square-foot food store, and 35,300 square feet of retail and office space. The exterior masonry and concrete work has been refurbished, a new roof constructed, and double-glazed windows added to the building exterior. Two large six-storey openings have been cut into the building to bring fresh air and light into the building.

The residential lofts are accessed by shuttle elevators from grade. Parking is on the second and third floors. On the fourth floor there is a lobby and recreation area consisting of a fitness club, basketball court, steam and sauna facilities, concierge station, two guest hotel suites, and a large area for social gatherings.

The loft apartments are divided into three condominium communities, each having its own elevators, stairs, and mechanical and electrical systems. The apartments have 12-foot ceilings and large expanses of glass. Many of the apartments feature in their design the massive (48 inch diameter) interior concrete structural columns.

The three condominium communities share a rooftop "sky garden" with prairie meadow flowerbeds and a wetland garden. A new twelfth floor has been added to make space for an indoor pool, party room, dining facilities, and an outside dipping pool.

The total redesign and regeneration of this 1,070,000-square-foot complex is believed to be the largest of its kind in North America.

Paul Northgrave

Sherbourne Lanes

241–285 Sherbourne Street
Completed 1976
Architects, A. J. Diamond and Barton Myers

Sherbourne Lanes stands out as a defining example of the impact of Toronto's reform movement on the city's architecture. The project represents an innovative way of providing high density housing in downtown Toronto. Following an era when most development involved the demolition of substantial tracts of the city's traditional Victorian fabric to make way for apartment towers, this scheme proved that a high density residential project could be care-

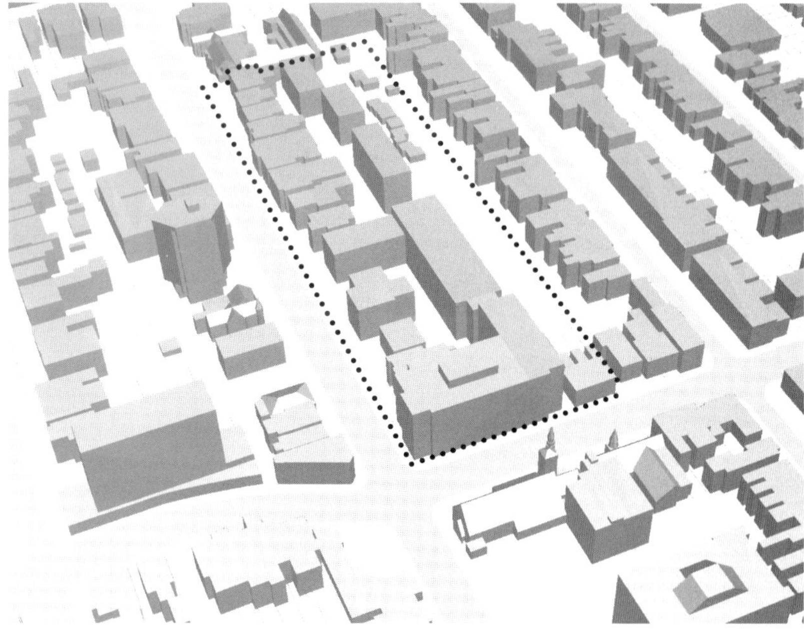

CPD/CT

fully inserted without disrupting the scale, material qualities, and urban virtues of the existing fabric. A city block of substantial Victorian homes, originally slated for demolition to make way for two 28-storey towers, was saved when Diamond and Myers produced a scheme that provided as many units as the two towers and maintained most of the existing houses. These were renovated to accommodate multiple units, and a six-storey apartment block was discreetly inserted at the rear of the houses' original deep lots, preserving Sherbourne Street's historic character. This was the first project by the City's new non-profit housing company, CityHome, and it represented an important watershed in the City's approach to urban renewal. Sherbourne Lanes received a Heritage Canada National Honour Award, a City of Toronto Non-Profit Housing Corporation Award, and an Ontario Association of Architects Award of Excellence.

Marco Polo

The Four Corners: epicentre of the housing crisis

The intersection of Sherbourne and Dundas could be called the epicentre of the homeless disaster and housing crisis in Toronto. Long known for its extremes of poverty, human conditions in the area of Sherbourne and Dundas are worsening.

Many things remain unchanged since I first "cut my teeth" as a street nurse in All Saints Church 13 years ago. Buildings are still boarded up, people wander the streets, and most of the social services are in their original locations. Yet, a closer look uncovers a real tragedy. Social service agencies report seeing double the number of people since the provincial government cut welfare payments, and health workers report two to four homeless deaths a week in the city. They also know this is a severe underestimation of the death count.

The Sherbourne and Dundas intersection is an essential part of what we call the Disaster Tour that we give to community leaders, politicians, and members of the media to show the extent of suffering from Toronto's housing crisis. First on the tour is All Saints, the magnificent Anglican Church at the Four Corners, which sports a bright new, blue sign listing the services inside: Street Health; a clothing store; the Open Door Drop-In; the Friendship Centre; and church service on Sunday at 11 am. Inside, in the Open Door, Doreen's 50-cent egg sandwiches are still my favourite in town. People that are homeless can seek daytime shelter in both the Friendship Centre and the Open Door. Today, I see still-homeless people I knew 13 years ago and many new faces too. Although people are now allowed to sleep on the pews, many are sleep deprived from nights in crowded emergency shelters or sleeping outside. In one neighbourhood shelter, for example, close to 120 people sleep side by side on a concrete floor. No cots, sometimes not even enough blankets.

Across the street from All Saints, bold signs scream "ROOM FOR RENT." These rooming houses are in an extreme state of disrepair, yet the need for housing is so desperate now that the landlord often puts two people in single, unfurnished rooms and charges each $320.

Down the street at Sherbourne and Queen there is an emergency shelter for men. North of All Saints, yellow brick houses remain boarded up, but sport a sign "Toronto needs new Public Housing Now." Next to the boarded-up housing, there is a "rest home" that offers minimal rest, minimal health services, but a bed. The cost is high.

At Allan Gardens there are visible changes from 13 years ago. New lighting and other design changes discourage public sleeping or loitering with limited success. Allan Gardens is now colloquially known as the "Safe Park" after the Ontario Coalition Against Poverty's attempt to create a safe

Michael McClelland

place where homeless people could sleep. Students from the University of Toronto continue to sleep there every Friday night in a public appeal for solidarity with homeless people.

Across the street women and children sleep in a permanent shelter. The majority of homeless women and children, however, have been relocated to emergency shelters in the former city of Scarborough. The count for children there is 1,000.

Juxtaposed with the hotel and condominiums sprouting up on Jarvis and Dundas streets, the Sherbourne and Dundas area has had no new social housing for 13 years. It's time to bring our people back into a home.

Cathy Crowe

The boom in temporary shelters

Overnight shelters in Toronto provide temporary accommodation for more than 65,000 people annually. That's almost double the number from 10 years ago. Children and families recorded the biggest increase. Almost 6,000 children are forced to live in homeless shelters each year. Some hostels are managed by the City of Toronto. Others are operated by private charities and receive government funding. Most provide dormitory-style accommodation – rows of bunk beds. Some families stay in suburban "welfare motels" where entire families are crowded into a single room. In the winter, some churches open their basements as part of "Out of the Cold" programs. Homeless people sleep on mats on the floor. The city's homelessness disaster means that on many nights of the year Toronto's homeless shelters are entirely full or at over capacity.

Michael Shapcott

All Saints Church, Homes for Tomorrow Society

319 Dundas Street East
Architects, Howard Walker and Howard Chapman
Completed 1988

All Saints Church is the subject of the book *My Parish is Revolting* [General Publishing Co. Ltd., 1974]. The book chronicles the transformation, under the supervision of Father Norman Ellis, of this Anglican parish church in the urban core, from a failing congregation in the 1960s and 1970s as more affluent members left for suburbia, to a revitalized Centre for the low-income people that were left behind. The revival of the church's presence in the urban core continued into the1980s under Father Brad Lennon's leadership. As the working-class neighbourhood was re-populated and renovated by affluent "white-painters," the issue switched to preserving housing for the Centre's low-income patrons who were threatened with displacement.

The Church's Board rose to the challenge. It engaged the groups using the Centre to plan for the whole site. The need for housing was on everyone's agenda. The Board convinced the Anglican Diocese of Toronto to apply to re-zone the land for housing, and to contribute the land to a project to build community space, offices for programs, and housing. The housing was financed with a provincial non-profit program.

EJR

The units were small but all self-contained, different from the rooming-house-based models elsewhere. Prospective tenants were involved in design meetings, and continued on as management committee members, cleaners, and repair workers.

The entrance in between the Church and the old parish house to the east (where programs are held), is obviously newer construction, but evokes the Church and parish house architecture. The housing has its own entrance, which serves as an unobtrusive backdrop to the historic church buildings, but is physically linked to the buildings on either side. The complex of buildings preserves the presence of the low-income community that has been based in the neighbourhood for over 100 years. The old and the new, the drop-ins and permanent housing, and the government, secular agencies, and the Anglican Church in partnership with the low-income community, make it a fascinating island of hope in the urban core.

Bill Bosworth

Robertson House Crisis Care Centre

119 Sherbourne Street
Current architects, Siamak Hariri, Taylor Hariri Pontarini Architects
Landscape Architects: MBTW Group
Completed 1998

Robertson House is a Metro Toronto Community Services temporary shelter for women and their children. Two connected historic houses on Sherbourne Street were renovated and a new L-shaped addition added. The facility is organized around a new central courtyard for social gatherings and children's play. The building attempts to avoid the "special housing" label by blending in with the scale of its urban surroundings. The image of the Robertson House addition draws on the metaphor of a single house, reinforcing the notion of a safe place of collective and individual living, which is both supportive and nurturing. The design attempts to be sensitive and understated, respecting the desires and needs of the users.

The three principal programmatic elements – the child-care facility, dining area, and residents' lounge area – open directly onto the landscaped courtyard, which has a small secluded garden at the west end. The presence of children is celebrated by a circular story-telling room, the indoor play area, and the separate youth activity room. Also on the ground floor are connected prayer, study, and counseling areas. The lobby, which is staffed by a receptionist gate-keeper is designed to accommodate the flow of residents, visitors, and baby carriages, and to provide a feeling of security from the street. Ground floor circulation occurs along the courtyard perimeter and incorporates spaces for casual conversation, repose, and encounter.

The bedrooms, all on the second floor, are modest and intimate in scale. Each bedroom contains a dormer window and window seat, providing a private, quiet outlook for mother and child.

Siamak Hariri

John Ross Robertson, 1841-1918

Publisher and philanthropist, John Ross Robertson lived in this house, 1881-1918. He was born in Toronto and while at Upper Canada College he started *The College Times*, the first school newspaper in Canada. He became city editor of *The Globe* in 1865 and the following year with James B. Cook established *The Daily Telegraph* published in 1872. Four years later Robertson founded *The Evening Telegram*, which quickly became one of Toronto's leading newspapers. Financial success enabled him to make substantial contributions to the building and operation of the Hospital for Sick Children and to gratify his life-long interest in history. He assembled an invaluable historical and pictorial collection and published some notable works such as "Landmarks of Toronto" and the "History of Freemasonry in Canada."

Plaque text courtesy of the Ontario Heritage Foundation

• Taylor Hariri Pontarini

Regent Park

Bounded by Gerrard, Parliament, Shuter and River streets
Completed 1947–59

Regent Park was Canada's first, and remains Canada's largest, public housing project. It was built in two phases between 1947 and 1959. The project replaced a neighbourhood that was similar to the one that remains immediately to the north, however, unlike the adjacent, now gentrified, "Cabbagetown," the Regent Park neighbourhood had been declared a slum. Following the analysis of "environmental determinism" common at the time, the poor physical conditions of the neighbourhood were seen as the root of the social problems experienced by the residents. The urban renewal of the area followed the edicts of modern architecture, setting buildings in park-like surroundings, segregating pedestrian and vehicular traffic, and providing an architecture that assumed a set of universal needs on the part of the tenants.

EJR

The problems that have emerged in the neighbourhood have followed patterns similar to those in other North American public-housing neighbourhoods. Residents identify personal safety as an issue, noting that police have trouble patrolling a neighbourhood with no through streets. And, informal surveillance is also difficult because of the ambiguity about who controls the various semi-public spaces in Regent Park.

Originally intended to re-house residents displaced in the demolition of the old neighbourhood, Regent Park was designed with a culturally homogeneous population in mind. In the late 1940s, Toronto accommodated largely European, English-speaking communities. This has changed dramatically and social service agencies report that area residents now speak more than 60 languages. Regent Park has become a primary immigrant reception area, and in the process, one of Canada's most culturally diverse neighbourhoods.

Although it is common for outsiders to perceive Regent Park as an undesirable neighbourhood, the area includes many strong communities and dedicated residents who strive to improve the physical and social conditions, and to improve the neighbourhood's image in the eyes of the broader public.

Regent Park North

Completed in 1954, North Regent consists primarily of unadorned brick buildings in three-storey "dumbbell" and six-storey "dog-bone" configurations. Designed by architect J.E. Hoare, the designs mirrored other North American public-housing projects of the period. Most of the 1,200 apartments are designed for households with children, including a large proportion of five-bedroom units. Although all the through streets were closed, their pattern is still discernible and the buildings roughly address the old alignments. Pedestrianized Oak Street, along with central ball diamonds and swimming pool, constitute the main social spaces for the neighbourhood.

There is little connection between the units and the common exterior spaces. Residents and social service agencies have been addressing some of the safety issues associated with this problem by initiating popular community garden projects in some of the more problematic areas.

In the year 2000, the future of the neighbourhood is uncertain as the Conservative provincial government has "downloaded" responsibility for public housing to municipalities, although most cities cannot support such capital-intensive projects on the property tax base. Through the 1990s, there were a number of proposals to rebuild parts of neighbourhood, including options for private initiatives that would retain the same number of rent-geared-to-income apartments while introducing a mix of household incomes.

Regent Park South

The completion of South Regent in 1959 was accompanied by great praise from the architectural community. The neighbourhood is dominated by five towers designed by Peter Dickinson of Page and Steele Architects and townhouses designed by J.E. Hoare. The project won the Massey Medal for Architecture and was cited for its innovation. The 14-storey buildings, which are oriented to the points of the compass rather than the street grid, define the major public space, Saints' Square. The buildings themselves have a "skip-stop" elevator system that allows units to have two aspects and through-ventilation. Like North Regent, most of the units, even in the towers, were designed for families with children.

The 1959 *Canadian Architect* review of the project comments on the tower facades: "Envisaged as patterned walls ... their vibrant pattern of glass and brick firmly controlled by the grid of the structure. The central recession produces an upward swing of the two side wings which, where the escape balconies punctuate the facade, has a vivacity that fills the court below with the joy of rhythm." Residents, especially those with children, find the buildings problematic, however, and many prefer to live in the more banal buildings of North Regent.

Richard Milgrom

Regent Park South: Maisonette Towers

2–52 Blevins Terrace and
1–73 Belshaw Place
Architects, Peter Dickinson with Page and Steele
Completed 1958

As the Toronto slums were being bulldozed in the 1950s, the City entered into a controversial period of public housing and modern planning. Peter Dickinson's maisonette towers, located just south of Shuter, on Blevins and Belshaw, formed part of the second phase of the Regent Park redevelopment which was well under way by 1955. Regent Park was seen as a community that would be protected from the old, unredeeming slums of nearby Cabbagetown and, to signify a break with this past, Dickinson based his design on modernist planning set forth by Le Corbusier and others. Dickinson's towers for Regent Park South were executed while he was chief design architect with Page and Steele in Toronto.

The project is comprised of five 14-storey apartment towers sited around a central park known as Saints' Square. The chief planner for the project, Ian Maclennan, abandoned through streets in favour of dead-end streets. Walking around the site today, the towers seem to be sited randomly with no relationship to Dundas or Shuter Street, or the edges of Saints' Square. One does notice, however, that they are aligned carefully with the path of the sun, although at ground level, the pedestrian is barely aware of this fact.

Dickinson won national recognition for these towers when he received the prestigious Massey Silver Medal in 1958. He was proud of the open spaces created around the towers. On paper, the spaces looked attractive, but in reality, they were unadorned and empty of life. The inconvenience of poor shopping facilities and poor taxi, ambulance, and truck delivery access further isolated these towers and their communities, despite the proximity to a major thoroughfare. At Belshaw Place, for example, three or four token services are meant to satisfy several towers. Next to these insufficient services, Dickinson designed a heroic glass curtain wall exposing the mechanical plant on the ground floor.

Influenced by Le Corbusier, the interior design of the towers was innovative for its use of hallways on every other floor, allowing apartments to be built on two floors, with an internal stair connection. Le Corbusier first used this concept in the *Unité d'habitation* in Marseilles. The large apartments had at least two bedrooms that would work well to accommodate the population boom in post-war Canada. One curious design feature is the small balconies that give exterior access to adjoining units. In the event of a fire, one exits to the common balcony, then through their neighbour's apartment. In case the neighbour was away or the door locked, Dickinson designed a glass box to house a hammer that could be used to break into the neighbour's apartment, so an escape could be made.

Despite faults, Dickinson's maisonette towers are seminal buildings in the history of public housing in Toronto.

Ian Chodikoff

EJR

Trefann Court

Bounded by Parliament, Shuter, River and Queen streets
Rehabilitated 1971, 1973, 1978

In the first years of the Depression, the Lieutenant Governor of Ontario, Herbert Bruce, gave a stirring speech about poverty and grim housing conditions in Toronto. He urged that most central Toronto neighbourhoods be demolished to make way for new housing. After the Second World War there was a strong push to fulfill Bruce's dream.

Neighbourhood demolition and reconstruction was called "urban renewal" and it had an active life in downtown Toronto from the late 1940s to the late 1960s. At that point residents in the Trefann Court area raised such political heat that the urban renewal strategy was halted in Toronto and in other Canadian cities and new solutions for regeneration were found.

The Trefann Court area is bounded by Queen, Parliament, Shuter and River streets. The City proposed to demolish almost all of the residential and commercial structures in the area; to build public housing in the westerly portion of the site, complimenting the Regent Park urban renewal area immediately to the north; and to permit industrial buildings on the east of the site.

With the help of community workers, the residents urged the City that their homes not be expropriated, that they be allowed to have significant involvement in replanning their

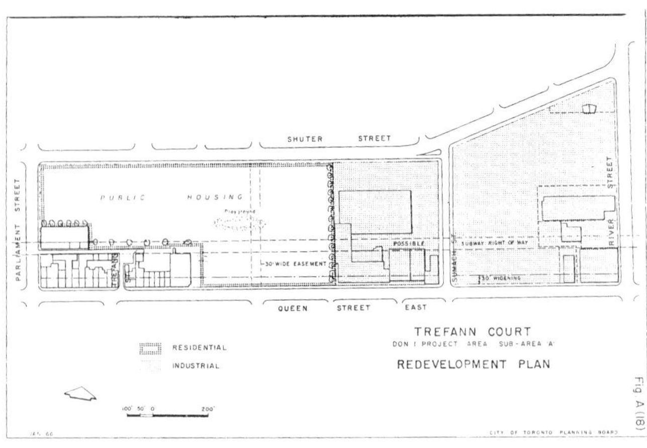

• *Sewell*, The Shape of the City

community, and that an alternative to public housing be found. By 1971 the residents had made significant changes in the way the City looked at the Trefann neighbourhood and other downtown communities.

The following changes were significant:

- A planner was hired by the City to work from a neighbourhood office directly with residents. This was the first example in Toronto of citizen participation in planning, a policy quickly adopted by City Council for all neighbourhoods.
- The residents and the planner agreed that most of the existing community should be protected, rather than demolished, and that new construction should strengthen, rather than weaken, the existing community. This was the first example of the policy to preserve neighbourhoods.
- An alternative was found to public housing (where every household had a low income and paid 25 per cent of income in rent). In Trefann Court, a small housing project, of about 30 units, provided a mix of home ownership, market rental, and subsidized rental. This approach of mixing tenures and incomes was quickly repeated throughout Toronto, most notably in the St Lawrence community.
- Trefann residents convinced the provincial government to enact a new Expropriation Law that included the principle "home for a home." The new law required an expropriating authority to pay full replacement value, rather than simple market value that reflected depressed prices because of urban renewal designation. The difference in cost to the expropriating body has been so significant that since this legislation was passed in the early 1970s there have been very few expropriations in Ontario.

The Trefann community barely stands out from other downtown neighbourhoods – which is one of its successes. However, one can find the innovative mixed-tenure, non-profit, housing project on the west side of Trefann Street (the first street east of Parliament); new City of Toronto Non-Profit Housing Corporation units on Sackville Street; and through the rest of the area, a mix of new and old housing, private and non-profit, some buildings more attractive than others. On Sumach Street, between Queen and Shuter, there are the remnants of the only large, new industrial building in the area that the City subsidized as a prelude to urban renewal, but it has now been converted to condominiums.

EJR

(For further information, see John Sewell, The Shape of the City: Toronto Struggles with Modern Planning. *University of Toronto Press, 1993.)*

John Sewell.

Toronto Women's Housing Co-operative

397 Shuter Street
Architect, Philip Goldsmith, Quadrangle Architects
Completed 1984

On the south side of Shuter Street, west of Parliament there are two matching rowhouse buildings, half a block apart, one perpendicular and one parallel to the street. This is the Toronto Women's Housing Co-operative (TWHC), the second of three women's non-profit co-operatives built in Toronto. The Constance Hamilton Housing Co-operative, located in the Frankel-Lambert neighbourhood, opened in 1982, and the OWN (Older Women's Network) Housing Co-operative, located south of St Lawrence Market, opened in 1997.

Incorporated in 1984, TWHC is also called the Beguinage by founding resident-members who were inspired by a historic model of women's communities – medieval European beguinages were small urban communities of unmarried women who lived apart from conventional families and at arm's length from male-dominated church control.

EJR

The stacked townhouse design of TWHC contains 28 units, with 13 one-bedroom units, 12 two-bedroom units, and 3 three-bedroom units. Each unit has a large private balcony or a fenced yard at ground level. Townhouse units have direct access to the street, and one-bedroom units share small vestibules. The project includes a co-ordinator's office, a workshop or storage room, and a laundry room. Due to design constraints imposed by the government funding program and because the TWHC was a turnkey development, there is little in the building design or layout to distinguish women's ideas and preferences, other than the communal courtyards at the back. Though not articulated in its physical form, the TWHC is distinctive in its creation of a unique women's space and supportive community.

The TWHC is an example of how women's groups across Canada have used existing social housing programs to develop non-profit housing for women. The self-management model of housing co-operatives has provided opportunities for residents to develop their management and leadership skills as well as to maintain control of their residential environment and provide mutual support.

Sylvia Novac

61 Seaton Street

Architect, Paul Reuber
Expected completion, 2000

The row house at 63 Seaton Street was originally situated in the middle of a block of attached Victorian terrace houses. In 1954, Shuter Street, which at the time ended at Sherbourne Street to the west, was extended eastward across Seaton to Parliament Street. As a result, the row houses south of 63 Seaton were demolished and the south party wall of 63 was suddenly exposed to Shuter. Although the street frontage of the wedge-shaped side yard formed to the south of 63 Seaton is only 4 centimetres, the rear yard's dimension of 7.8 metres was regarded by the owner as decidedly more tempting...

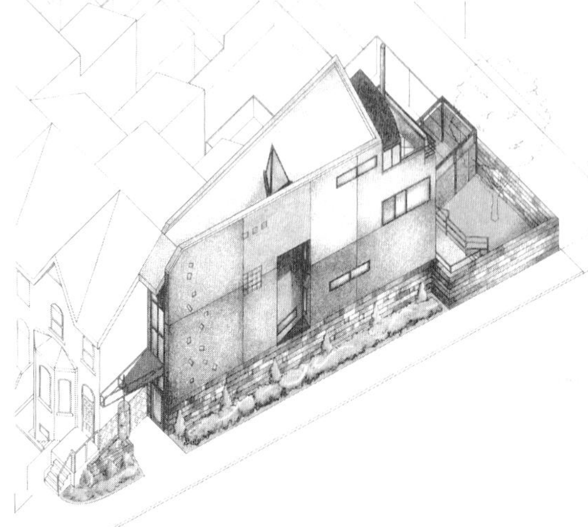

• Paul Reuber

In 1989, a new three-storey infill house was designed for the wedge-shaped side yard. The design for the new free-hold house accommodated the specific living requirements of the owner of 63 Seaton – a doctor with an extensive art collection. The south end wall of 63 Seaton has been restored to its former party-wall status. Owner and architect wanted the new building to *cauterize* a part of the city block that had been previously *wounded* by the rather barbarian insertion of a new roadway.

Although this house is shaped like a piece of pie, City Hall approval to build it was no piece of cake. Neither the neighbours, nor the Planning Department shared the enthusiasm of the owner and architect. Fortunately, the Ontario Municipal Board approved its construction after a time-consuming and costly appeal. Then the capriciousness of the housing market and the owner's purchase of a new home elsewhere nearly caused the project to be shelved permanently; however, it was extensively redesigned and built as a speculative venture eleven years after its initial conception.

The front of the house may be only an entry door wide, but the principal rooms at the rear are more spacious and airy than those in its Victorian neighbours.

Paul Reuber

Proposed redevelopment of Moss Park Apartments

Bounded by Sherbourne, Shuter, Parliament and Queen streets
Architects, Somerville McMurrich & Oxley, with Gibson & Pokorny, and Wilson & Newton
Completed 1961

In the early 1960s, the City razed several blocks of 19th century Victorian housing and constructed three 16-storey, Y-shaped, public housing towers known as Moss Park Apartments. In the 1970s, a fourth tower was added. Under-utilized parking lots partially surround these towers at grade. To the south a sizable park is well used by tenants and the local community.

In 1991, the Province of Ontario's Ministry of Housing funded a community-consulting process called the Moss Park Community Development Project. A working committee was formed that included building residents, the City, Homes First Society and the Supportive Housing Coalition of Toronto (both providers of non-profit housing), and the writer as the committee's planning consultant. Two phases of work were identified: Phase I calls for the refurbishing of existing housing towers; Phase II proposes infilling the super-block's parking lots with 220 new, affordable housing units, which would eventually pay for Phase I.

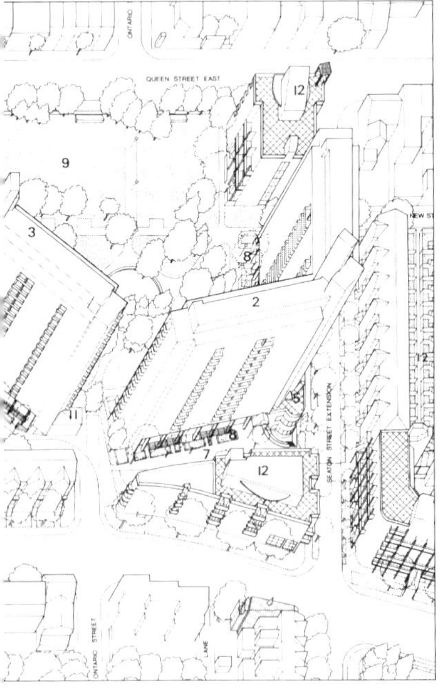

The Urban Design Guidelines set out in Phase I have, in part, been executed by architect Ted Sievenpiper and include new lobbies and amenity spaces.

Phase II calls for a new system of public streets, lanes, and pedestrian strollways to create eight new building sites. Different housing typologies were developed to respond to the particular size, shape, and location of each site, and include row houses, stacked row houses, several smaller apartment houses, and a larger courtyard apartment block. The plan sets a community/cultural centre in the park to serve the entire neighbourhood. It was proposed that families, currently living on the upper floors of the deteriorating towers, be given first choice to move into the new housing at grade. As older units became vacant, they were to be refurbished to house adults on marginal incomes.

The partial implementation of Phase I enjoys a somewhat peculiar relationship to the existing site plan due to the cancellation of Phase II, which remains a theoretical planning "vision."

Paul Reuber

● *Paul Reuber*

Moss Park

Bounded by Jarvis, Shuter, Sherbourne and Queen streets
Built c. 1830; demolished

Moss Park, with its baseball diamonds and recreation centre, is perhaps hard to imagine as the estate grounds of one of Toronto's early family dynasties – the family of William Allan (1772–1853). The park is notable for the following reasons: 1) Moss Park and, further north, Allan Gardens, are one of the few *green* reminders of the park lot system that subdivided early Toronto into its aristocratic parts; 2) These park lots, running from present-day Queen Street to Bloor Street and defined roughly by today's north-south streets (in William Allan's case, George Street to Sherbourne Street), defined the cartography by which Toronto housing lots would develop; 3) Moss Park is a reminder of the Allan family's interest in promoting the civic virtues of landscaping and horticulture.

A handsome Greek Revival house set in a picturesque landscape, the northern reaches of the Moss Park estate were subdivided in 1850.Tree-lined Pembroke Street was built, and stately homes marched north. The Allans maintained their estate house in Moss Park for some time. Allan Gardens was donated in 1860 to the Toronto Horticultural Society by William Allan's son, George (the future Mayor of the city), "for the enjoyment of all citizens." It was meant to be surrounded by a Nash-like oval, divided with lots for villas, but this idea was never realized.

David Winterton

90 Shuter Street

Architects, Tsow and Pollard
Completed 1984

The building at 90 Shuter Street is the Homes First Society's first project to replace the low-rent housing and rooming house stock that was lost in the late 1970s real estate boom. Purpose-built, it has 17 "rooming houses," i.e., 17 four- and five-bedroom apartments, two per floor in the 11-storey apartment building. The rooms are large. Drywalled concrete block walls provide sound and fire separation.

The design maximizes individual privacy. There are only four or five people per apartment and only nine people per floor. The small groups facilitate group interaction, decision making, and problem solving. Tenants have keys to stop the elevator only on their floor, to promote the sense that the apartment-front-door is the door-to-the-street.

The lounge in elevator lobbies on each floor is for interactions between the two apartment groups. The ground-floor common space (with kitchen), by the front door, is for "drop-in" relationships on the way in or out, parties, and programs. The second floor common room is for organized meetings, parties, and programs. The top-floor deck provides outdoor space away from "friends" on the street. A top-floor workshop space is available for hobbies and for community economic development. Offices on the ground floor are for staff and tenant management, and agency service delivery.

The partnership of community agencies, citizens, tenants, and staff at 90 Shuter represents a holistic, community-development approach. Tenants self-selected into the initial groups, in a unique organizing process that allowed them to develop expectations and group rules before occupancy. Community members from the Board sat on committees with tenants and staff to ensure due process, to help solve problems, and to consider evictions.

In the year 2000, in a mean-spirited political climate where "the bottom-line" rules, this holistic philosophy has a low standing in many people's eyes. Yet, when you see a former "biker" chide an agency worker for marking up the paint on moving-day – "Hey!! This is my house!!" – the benefits of the approach are obvious.

Bill Bosworth

Fred Victor Centre – Keith Whitney Homes

147 Queen Street East
Van Nostrand Di Castri Architects
Completed (renovations) 1990

The Queen and Jarvis building dates from the 1950s and sits on the site donated by the elder Hart Massey in the late 1800s. The building was constructed as a men's hostel and senior men's home, the vision of Reverend Wesley Hunnisett, who had housed transient working men in the St Lawrence Market during the Depression (the same man and his philosophy was behind Seaton House, the City hostel on George Street).

Fred Victor Mission has served low-income people for over one hundred years. It started as a mission of the "Methodist Cathedral" (now the Metropolitan United Church), it ran schools, penny savings banks, fuel co-operatives, and programs for families in the low-income neighbourhood to the north and east.

Before renovation by John van Nostrand, the building was an industrial-looking, dull-yellow brick, with aluminum windows, built to the lot lines on Queen and Jarvis streets. It was home to 120 men in the hostel each night, and 60 senior men in the home. The hostel residents were "permanent," just on a rotation in and out for a few days, months, or years at a time.

The goal of the project was to replace the 180 "beds" on the site. The architect faced the challenges with creativity. Since setback requirements for new construction would have limited the potential of the site, the building had to be renovated. He inserted reverse-bays into the walls to bring light into the new multi-room units. The cladding lightened the impression of building mass despite the fact that the "wrong colour" was delivered in the midst of the building boom and could not be corrected.

The Fred Victor "Mission" became the "Centre" with computer training, support for economic development, a cafeteria, and many innovative programs to support the lives of the men and women in the housing, and those still on the streets, who see Fred Victor as their "home" if not their residence.

Bill Bosworth

Rooming houses

Beginning in the 1960s and early 1970s new provincial policies encouraged the deinstitutionalization of people with mental illness, which resulted in an increased demand for affordable housing. At the same time, the supply of affordable housing stock in Toronto diminished because of gentrification in the downtown areas (especially of rooming houses), low rental vacancy rates, and, finally, high unemployment forced many people to seek cheaper housing. Thus, as early as the 1980s, Toronto began to experience a crisis relating to the lack of affordable housing.

Rooming and boarding houses form an essential component of the housing continuum since they provide low-cost housing at the very bottom of Toronto's housing market. For many vulnerable people (often mentally ill and not receiving treatment) the only other choice is the street. Unlike self-contained housing, rooming and boarding houses have at least one shared facility: bathroom, kitchen, or living room. Rooming houses provide accommodation only; boarding homes provide meals and care services.

In 1974 Toronto City Council passed by-laws to require owners of non-owner-occupied rooming houses with five or more tenants to obtain licenses, submit to inspections by City officials, and meet standards for fire protection and maintenance. It has been argued that this regulation has been a major factor in the significant decline of the stock; however, decline was also fueled by gentrification and other forms of urban development in the mid-1970s. Rooming houses have been regulated since 1987 by the Tenant Protection Act. From 1986 to July 1999, licensed rooming house stock has declined from about 600 homes to 389 homes. Boarding home stock has always been more limited. Of the rooming houses currently licensed in the former City of Toronto, only 60 operate as boarding homes. Currently, rooming and boarding houses are only legal within the old boundaries of the City of Toronto and Etobicoke. Most are located in Parkdale and Cabbagetown, close to the community services that rooming and boarding house tenants rely on. Neighbourhood opposition will make it difficult for existing licensing to be extended to the other four former municipalities that were amalgamated to create the new "megacity" of Toronto.

Agencies working with the homeless see other pressures further reducing the boarding and rooming house stock. The province, for example, has removed its legislation protecting rental housing, opening the way for the conversion of rooming houses to single family housing, rental apartments, or condominiums. And the removal of rent controls has permitted rents to escalate. The typical monthly rent for a decent room in a rooming house in Toronto has gone from $425 to $500 over the last two years, well above what people on social incomes are able to pay. Changing municipal taxation policy may quadruple the taxes for larger rooming houses, thereby raising rents; and the evolution and enforcement of municipal standards for rooming and boarding houses may also raise owner's costs. Some owners and operators are now reaching retirement age and are looking to sell their properties and/or the business.

Trends in the banking and insurance businesses are also increasing the costs of operating boarding and rooming homes. Trust companies, the usual mortgagor of the houses, are being absorbed by the major banks, who subsequently withdraw from this line of investment. Rooming house owners are forced to seek private or offshore investors for mortgage financing at significantly higher interest rates. Insurance companies are also scrutinizing the boarding home business, and charging higher premiums.

The housing market is becoming tighter. In the private sector it is easier to increase rents, forcing those who could afford previously to live in self-contained apartments to move into rooming or boarding homes. Landlords of rooming and boarding homes have become more selective in choosing their tenants, and current tenants are being forced out and into substandard housing or onto the street.

The City of Toronto must develop strategies to prevent the loss of valuable rooming and boarding house stock and to encourage new investors into the sector.

Alison Guyton

The Rupert pilot project

Rooming houses (sometimes called Single Room Occupancy units) provide affordable housing for low-income people. Since the 1970s, gentrification has led to the loss of tens of thousands of rooms, which have been replaced by high-priced "yuppie palaces." The migration of young professionals into poor neighbourhoods has also led to demands for more aggressive policing against low-income and street people. A fire in December 1989 at the Rupert Hotel at Queen and Parliament streets, which left 10 residents dead, sparked an innovative pilot project. Co-ordinated by a community-based group, the Rupert pilot project used provincial and municipal funds to renovate more than 500 units of rundown rooming house stock. The project created successful models for collaboration between government and community, but new development stopped in 1993 when the provincial government refused to renew funding.

Michael Shapcott

The Derby

393 King Street East
Architect, Dermot J. Sweeny
Completed 1988

The Derby, which stands at the southeast corner of Parliament and King, makes a handsome terminus for the King Street vista seen from the west. The project, initiated by Ron Thom's

EJR

Toronto office, was brought to a successful conclusion by Dermot Sweeny. Twenty-six innovative mezzanine lofts (long before "loft" became the ultimate marketing buzz-word it is today) and the commercial space at street level fit in well in this zone of (formerly) clandestine lofts and industrial buildings.

David Winterton

Live/work – a personal memoir

Northeast corner of King and Parliament streets

It was 1978. A newly registered architect, with more ambition than cash flow, was looking to open a new office. The City of Toronto was, against enormous pressure, pursuing a policy of preserving industrially zoned land. The over-heated real estate market had created absurdly high rental and purchase costs for traditional housing.

At the corner of King and Parliament, a landlord combined with a small group of prospective tenants to create a loft community in a downtown warehouse building. It was one of a number of illegal lofts that were springing up on the shoulders of Toronto's downtown, originated by people who were following the example of artists and others in New York, London and elsewhere around the world, and who were willing to take risks in return for unique opportunities.

The building at King and Parliament offered brick walls, mill construction wood floors, high ceilings, huge windows with great views, and total flexibility in layout. It also featured substandard heating, freight elevators that were an adventure with every use, and a virtually total disregard for fire safety codes. But for a new architectural practice, the $250 per month rent (and that was gross rent!) for 750 square feet of "funky" space was an offer one could not refuse. The space, at the northeast corner of the fourth floor, subdivided easily into living and working areas, and suited the needs of a fledgling practice. Working hours were flexible, and the area lent itself well to entertaining or business meetings, whether long solitary hours of drafting, groups pulling an all-nighter, dinner for two, or a working lunch.

It was presentable as a business address and met the desire for a downtown lifestyle, being within a few minutes walk of King and Yonge, the St Lawrence Market, restaurants, theatres, and the squash club. The daily experience of watching the downtown-bound traffic pour onto Richmond Street from the Don Valley Parkway as I dressed for the day reinforced the conviction that this was a viable alternative to the suburban/urban dichotomy that had become the norm in North America.

The City never made a serious attempt to eliminate the illegal residences. Perhaps the number of buildings and residents involved was so small that it would not have been worth the effort. Perhaps they recognized that, although the lofts did fly in the face of the City's industrial policy, there were benefits accruing to the City, such as the preservation of older industrial buildings, the provision of alternative forms of downtown housing, and the reinforcement of economic activity around the urban core.

The experiment continued for over two years, until the difficulty of making the transition from living to working became harder to achieve. As the staff complement of the office grew and their working schedule started to follow a more regular pattern, the acceptability of their being greeted at 9:00 a.m. by the principal in a bathrobe became less defensible. Bowing to the inevitable, the space became merely an illegal office in an industrially zoned building.

The King-Parliament experiments, and others like it in the King-Spadina area, presaged by two decades, the introduction of the Revitalization Area zoning during the tenure of Mayor Barbara Hall. In these areas ("the Kings"), the zoning does not proscribe uses or densities, but rather allows the market to determine appropriate uses. The zoning requires compli-

E/R

ance with building-envelope criteria and reduced parking and loading standards. Many previously empty and deteriorating structures have been rehabilitated and given new life. There is a new vitality in both districts, with new lofts and condominiums, high-tech offices, restaurants, entertainment establishments, light manufacturing, and the traditional fashion industry coexisting in a vigorous and dynamic synergy.

Leslie M. Klein

Corktown

Corktown is one of the city's oldest areas, characterized by housing nestled in between factories, breweries, and large industrial properties. Settled mainly by Irish immigrants, who came here after the Famine of 1846–47, it was dubbed "Cork," "Paddy," or "Slab" Town.

The needs of its residents have always been largely satisfied within the community. There are stores (the first Loblaws family store was located on King, east of Parliament), churches (Little Trinity and St Paul's Basilica), schools (Enoch Turner Schoolhouse on Trinity, which was Toronto's first free school for children, and the Inglenook Community High School on Sackville), pubs, and once an elegant performance space at the top of the Dominion Hotel and a local swimming hole at the Don River.

Corktown is a fascinating combination of very old row housing (up to 150 years), and the very latest conversions of industrial buildings (for example, 90 Sumach Street, now high-tech, loft condominiums, formerly a CBC sets-and-props location). Most of the housing that you see today on Bright, Percy, Trinity, and Ashby Place was built in the mid to late 1800s for families of local workers. Generally, it consists of narrow, two-storey row houses with dirt-floor basements, a brick front and wood rear walls finishing off a summer kitchen.

Bright Street, which curves gently between Queen and King, has a name that belies a history of poverty, illness (due to rising damp from an underground creek), and hardship.

In 1950, one hundred years after the first immigration from Ireland, the street was still populated by Irish families. According to a local resident, 125 children lived on the street when she grew up there – one family alone had 24 children.

Queen, King and Sumach streets (from Shuter to Queen) have larger-scale terrace homes, built in rows of four or five. The greater width, interior ceiling heights, and level of detailing were enjoyed by business managers and their families. Less imposing and less costly houses were built for skilled labourers. The home at 473 Queen Street East, for example, was built for a brewery manager by a noted local contractor, Mr Davies, who built the Dominion Brewery, directly across the street. There are five houses in this row, Davies Terrace, erected in 1877. The two remaining row houses to the east were also built on the same scale. The original occupant of the 15-foot-wide home at 495 Queen Street East, built in 1892, was James Lee, a typesetter. The western end of the block (Queen and Bright) peters down to a scale more in keeping with the row houses on Bright Street.

Hélène St Jacques

Bright Street

This small corner of Victorian Toronto is easily overlooked amidst the freeway ramps, auto shops, and assorted stores of lower Cabbagetown. But if you head east on Queen Street, and turn right a block or two before Queen crosses the Don, you will find yourself on Bright Street, a short, bending lane running south to King.

Small scale and lined with narrow row houses, Bright Street is peppered with tiny gardens and potted plants, chained bicycles, gates of various kinds, and bits of rubbish awaiting collection. The general effect is unexpected and haphazardly charming. In recent years Bright Street has attracted film location scouts seeking the old, the genuine, and the vaguely trans-Atlantic.

E/R

The street we see today is, in fact, mostly 19th century. Apart from a short row of brick and mansard-roofed cottages built in 1901, it is made up of groups of two-storey gabled houses from the 1870s. These were the homes of shoemakers, carpenters, and labourers who worked in neighbourhood industries such as the Gooderham and Worts distillery.

Now these houses are occupied by writers, students, real estate brokers, clerks, cartoonists, and musical agents. Despite the ongoing revival of the surrounding area as a desirable zone of lofts and studios for nouveau-urbanites, Bright Street seems to change little. A prominent sign still forbids ball playing in the street. A pocket park on the west side still attracts workers at lunch, children after school, teenagers at dusk, and down-and-outs pretty much any time at all. Bright Street is short of space, short of parking, sometimes noisy, and not particularly beautiful, but it is still interesting and attractive and it reminds us of a kind of ordinary urban life mostly gone from other parts of the city.

Kelly Crossman

West Don Lands

The West Don Lands is a largely disused, 80-acre parcel of land bounded by the Don River, Parliament Street, Eastern Avenue and the CN Railway embankment. Originally set aside in 1793 as a park reserve, it was gradually given over to residential and then industrial uses. As industry in the centre of Toronto declined, so did the fortunes of the West Don Lands and the surrounding neighbourhoods. In 1988, the entire site was expropriated to create a new residential neighbourhood called Ataritiri. By 1993, the development had been cancelled due to escalating development and environmental costs and a collapse in the real estate market. The Province took over the lands in 1996 and has been trying, unsuccessfully, to sell them ever since.

Rollo Myers

In 1997, the West Don Lands Committee, a 17-member coalition of local resident, business, environmental, and heritage organizations, was formed in response to reports that the lands were about to be sold for a harness-racing facility. Since that time the Committee has worked actively to redirect the development. In November 1999, the Committee hosted a development workshop that brought together the local residential and business community, environmental and heritage advocates, developers, municipal and provincial regulators, and design professionals to build a consensus about principles for development. Three multi-disciplinary design teams then prepared sample development scenarios based on those development principles.

The Toronto Waterfront Revitalization Task Force and the Toronto 2008 Olympic Bid Masterplan have also identified the West Don Lands and the long-neglected mouth of the Don River as a significant waterfront resource. Both initiatives have incorporated ideas and principles generated in the West Don Lands Committee workshop. Both reflect the community preference for a mixed-use, medium-density, residential neighbourhood, with a large park located on a flood-protection berm at the edge of the Don River. Both incorporate a significant project to re-naturalize the mouth and channel of the Don River, another important community priority.

Cynthia Wilkey

Three residential buildings in the Gooderham & Worts historic district

Bounded by Parliament, Mill and Cherry streets and the CN railway embankment
70 Mill Street and 39 Parliament Street: Quadrangle Architects
80 Mill Street: Northgrave Architect

Founded in 1832, Gooderham and Worts is a five-hectare parcel of 45 aging distillery buildings, declared a National Historic Site in 1988. A 20-minute walk to downtown, the site lies between the St Lawrence Neighbourhood and the "brown fields" of the West Don Lands.

As part of a 10-year mixed-use planning exercise, 875 residential units were assigned to the periphery of the site, 25% of which were to be "affordable" following the City guideline.

A change in provincial government saw an end to housing subsidies in the early 1990s. In response, Michel Labbé of *Options for Homes* devised a plan whereby co-op groups would be formed as non-profit developers to build condominium units. The units were priced as "affordable" under the City guideline, without any Government subsidies.

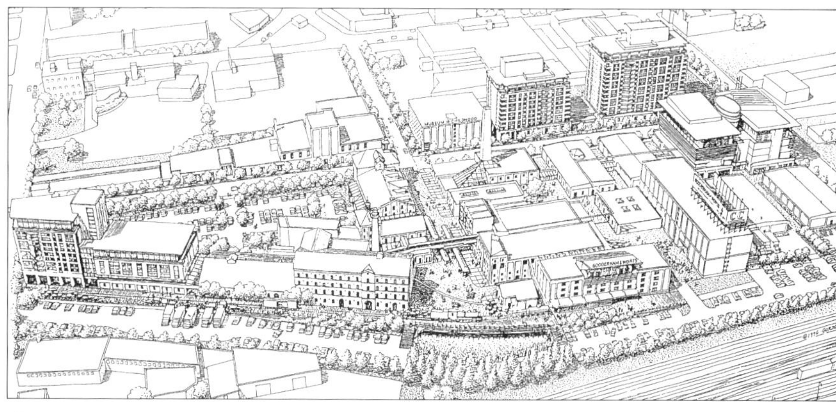

Gordon Grice c/o Davies Smith

To date, three buildings, containing 420 units in total, have been built under this plan. The balance of units will be market condominiums. The apartment towers rise out of 19th century rack and tank buildings on the edge of the site and overlook the historic Trinity Street core. The central area is slated for retail and office uses fitted into the heritage buildings.

All new buildings planned for the site are tightly controlled by built form guidelines that define setbacks, side yards, build-to lines, and building heights. These controls were put in place to ensure a harmonious relationship between the new development and the heritage building fabric.

David Dennis

St Lawrence Co-operative and Parliament Square

70 Mill Street
Quadrangle Architects Limited
Completed 1998

The historic Gooderham and Worts distillery complex is one of Toronto's oldest intact collections of historic buildings and is designated as a Historic District under federal legislation.

• Quadrangle

The master plan calls for the long-term development of a mixed-use community, including residential, office, retail, cultural, tourist, and entertainment uses. The first two residential buildings, located at 70 Mill Street and 39 Parliament Street were constructed in conformance with the master plan design guidelines. They have 95 and 177 units, respectively, and both incorporate portions of existing historic structures.

Leslie M. Klein

Market Square

80 Front Street East, 35 Church Street
Jerome Markson Architects Inc

Completed in two phases during the early 1980s, Market Square condominiums provide the historic St Lawrence area with a touch of Haussmann's 19th century Paris that deserves to be more widely emulated in Toronto. The project's overall density may be high (0.5 times the lot area is commercial, 4.5 times the lot area is residential), but a street-friendly, nine-storey height and a courtyard configuration define adjacent streets, lanes, and pedestrian ways with a solid perimeter of built form.

The complex includes: stores and restaurants on the first floor; six movie theatres and a commercial parking structure below grade; continuous commercial colonnades; and a mid-block pedestrian route that is strategically situated on the axis of one of Toronto's finest historic landmarks, St James' Cathedral. There are also generous bay windows on the upper floors that provide apartments with diagonal views up and down the street, and these, along with a roofscape of fireplace chimneys, outdoor terraces, and garret windows, create a residential imagery that is enhanced by the application of multi-coloured brick.

Fiona Spalding-Smith

Market Square is the home of several designers and architects partly due to its highly urbanized context, and partly because of its columnar structure that makes for easy suite renovation. As well, eighth floor suites have high ceilings and roof terraces commanding panoramic views of Toronto's skyline. A swimming pool with adjacent communal roof terrace is reserved on the ninth floor for the use of all residents.

Market Square was featured as an urban prototype in a study by the author titled "Shifting Gears for Fewer Gearshifts," sponsored by the Design Exchange. The study superimposed Market Square on five theoretical sites within Toronto's commercial core to demonstrate that the city's commercial and civic plazas (including Nathan Phillips Square), public streets, and underground PATH system have a tremendous ability to absorb more, high-density, low-rise, residential accommodation.

Paul Reuber

The St Lawrence Neighbourhood: a lesson for the future

Bounded by Yonge, Front, and Parliament streets, and the Canadian National railway embankment
Architects: Irving Grossman, Klein and Sears, Vaclav Kuchar and Associates, Boris Lebedinsky, Jerome Markson, Matsui Baer Vanstone Freeman, Robinson and Heinrichs, J.E. Sievenpiper, Sillaste and Nakashima, Thom Partnership
Completed between 1977–82

Undertaking a large-scale redevelopment in an inner-city area is rarely successful, particularly if one of the major objectives is creating housing for poor people. Urban regeneration, as the process is known, is even more suspect when undertaken by government. The St Lawrence Neighbourhood in downtown Toronto is a very successful and notable exception, and therefore of special interest.

Only six years before commencing the St Lawrence Neighbourhood, governments in Canada had publicly abandoned the process of *urban renewal* as an utter failure. The St Lawrence Neighbourhood happened only because the need for housing the poor remained urgent. It flowed from essentially the same set of objectives as urban renewal, but what made it different?

Perhaps the most important difference was a new set of housing programs created by the federal government in 1973, although a new approach to thinking about the design of such a project proved to be equally important to its success. The new programs, which were abandoned only 20 years later as being too expensive, went beyond the simple provision of basic housing for the poor. They encompassed social policy goals for assisting the poor to integrate into society and thereby became tools for dealing with wider urban issues from a social perspective.

These programs enabled the creation of co-operative and non-profit housing as a new way of housing the poor. Non-governmental organizations were able to provide housing for those in need, tenants could control and collectively own buildings, and most important of all, the programs mandated a mix of incomes. Old-style public housing had been directed only at those most in need, that is the poor often on welfare. The new programs would cater to a mix of the most needy and a proportion of working people in lower- and middle-income ranges.

In creating the St Lawrence Neighbourhood the objective was to avoid the *project* mentality of a group of buildings that produced the ghetto associated with urban renewal. Instead, the housing was to take its cue from the typical Toronto neighbourhood. From this starting point, three principles emerged: public streets should be the basis of the neighbourhood, buildings should address the streets, and there should be *mix*.

The typical Toronto neighbourhood is not very special, architecturally speaking – the houses are cookie-cutter, builder-designed, run-of-the-mill. But the individual houses are less important than their conglomeration along public streets, which are richly adorned with trees, and connect to schools, parks, and usually a main street with shops within a few blocks. The streets create the fabric that we know and love as the Toronto neighbourhood.

Streets in the St Lawrence Neighbourhood were planned to be just as in any other Toronto neighbourhood, that is they were to be public, continuous, and extended into and through the neighbourhood, integrating it fully with the framework of the surrounding city. This meant adding more streets rather than cutting off existing streets as was the norm in urban

MWF

renewal projects. For new buildings in St Lawrence, it was decreed that they would to relate directly to the public streets, with front doors opening onto the streets and with the public streets and lanes being the only place of pedestrian circulation – again a departure from the project mentality.

The final key principle was mandating *mix*, again a condition found in most neighbourhoods and never in projects. Mix in St Lawrence became almost a fetish. Development was to be mixed in housing type, housing tenure, income group, land use, and even developer. The neighbourhood was developed with a mix of non-profit, co-operative, and private owners, and among each there were many different developers. The non-profit and co-operative buildings all had tenants with a mix of incomes, made possible under the new housing program. A quarter or more of the housing was private market, primarily for ownership. The least successful was the mix of land uses; in the initial phases there could perhaps have been more retail and more work places.

St Lawrence features the best residential architecture produced in Toronto in the 1970s and 1980s. There are fine buildings by Henno Silaste, Irving Grossman, Ron Thom and Jerome Markson, among others. They worked with limited budgets and a narrow range of materials. They produced the first stacked townhouses in Toronto, now being built throughout Ontario, and their quality has yet to be exceeded. They also related higher apartment forms to rows of townhouses and the buildings worked in unison with interesting parking and open-space solutions. The earlier phases were all kept to red brick, unnecessarily abandoned in later phases over concerns about uniformity. Building heights were also allowed to creep up in the later phases. But neither the change in materials, the mix of fine architecture and ordinary buildings, nor somewhat oversized buildings seem to matter. Instead, they prove the importance of the simple principles that make it all work as successfully as the existing Toronto neighbourhoods.

Today, the neighbourhood extends well beyond the original boundaries. It is no longer apparent where the St Lawrence Neighbourhood project undertaken by the City of Toronto ends and where the rest of the city begins. Today, there is much more ownership housing than subsidized housing, and there is a considerable mix of land uses with much space for work, retail, entertainment and education. St Lawrence never was a project. It became a neighbourhood from the start, catering to many poor people, but also becoming a catalyst for the regeneration of a large part of the city.

In labelling the 1973 housing programs as too expensive, governments never acknowledged that the programs did more than just house the poor. Today, people are again demanding that governments house the poor and address questions of urban infrastructure. Let us hope that when governments are forced back into the housing business, they do not return to the mean-spirited model of urban renewal that is sure to fail, but look instead to the model that is best illustrated in the St Lawrence Neighbourhood. A made-in-Canada, proven solution – let's use it.

Frank Lewinberg

St Lawrence Neighbourhood Seniors Housing

85 The Esplanade
Quadrangle Architects Limited
Completed 1995

• *Quadrangle*

One of four non-profit projects built at the same time under the auspices of Cityhome, this development contains 128 units designed to assist the frail elderly to maintain independent lives. The buildings conform to built-form guidelines developed by the City and surround a common courtyard. Grade-related retail space and one level of underground parking are also provided.

Leslie M. Klein

C-2 Block, St Lawrence Neighbourhood

South side of the Esplanade,
between Market and Church streets
New Hibret Co-op:
Roger du Toit Architects
CityHome:
Stone Kohn McGuire Vogt Architects
Old York Tower:
Quadrangle Architects
Older Women's Network Co-op
(115 Esplanade):
Oleson Worland Architects
Completed c. 1995

The development of this block, at the west end of the St Lawrence neighbourhood, in a sense completes the first phase of one of North America's most successful "urban renewal" programs (population approxi mately 10,000). The St Lawrence Neighbourhood was planned by the City of Toronto approximately 20 years ago, as a new, mixed-use (primarily residential) neighbourhood, replacing an obsolete industrial area. The area has many amenities, including the historic St Lawrence Farmers' Market (which incorporates a portion of one of Toronto's early city halls), St Lawrence Centre theatre, and the Hummingbird Centre concert hall. It is within walking distance to the downtown core. The area is historic, being just west of the original 10 city blocks, with significant structures such as St Lawrence Hall, the Gooderham (Flatiron) Building, and the King Edward Hotel nearby.

The neighbourhood plan features a large, lineal park at its centre, running the full length, east-west, named Crombie Park after the mayor who initiated the project. The housing blocks, which were designed under a set of consistent urban design guidelines, are generally mid-rise (six to eight storeys), clad in brick, with convenience retail and pedestrian arcades at the base. The residential buildings are a mix of public (co-ops and City Housing) and private developments.

● Oleson Worland

The C-2 Block was laid out by the City to accommodate four residential buildings (three co-ops and a City Housing building), approximately 550 units total. The buildings ring the block around a central courtyard, one at each corner, and all above a common underground parking garage. Each building has a different owner, and was designed by a different architect: New Hibret Co-op (Roger du Toit Architects); CityHome (Stone Kohn McGuire Vogt Architects); Old York Tower (Quadrangle Architects); and Older Women's Network Co-op (Oleson Worland Architects).

Urban design guidelines were set for the block by Michael Spaziani Architect. Each architect designed the building for his client, within a shared, overall vision. Representatives of the four architecture firms comprised a committee to coordinate the project from beginning to end. All four buildings were clad in brick, with variations in expression and detailing. Landscape design of the central courtyard and streetscape edges was undertaken by du Toit Allsopp Hillier and Ferris and Quinn. It was agreed that there be one structural engineer, Frank Anrep and Associates, to ensure structural coordination, and all four buildings were tendered as one block to ensure coordination during construction. PCL Constructors was the general contractor.

The development of the C-2 block is important for several reasons: it marks the completion of the downtown end of the St Lawrence neighbourhood; it is an excellent example of diversity in architectural expression within a shared vision; and it is a prototype for future big-block developments in new neighbourhoods (for example, the Railway Lands, the East Don Lands, and the Portlands).

David Oleson

● *Oleson Worland*

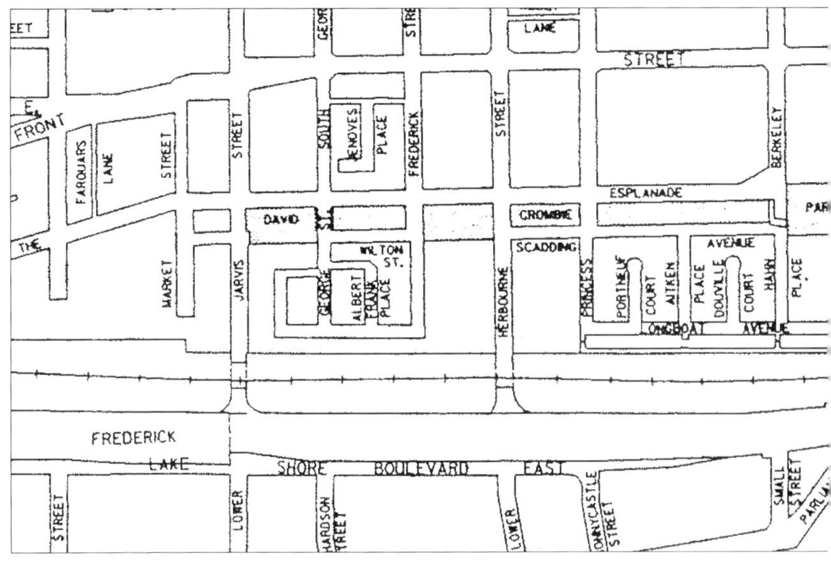

WEST

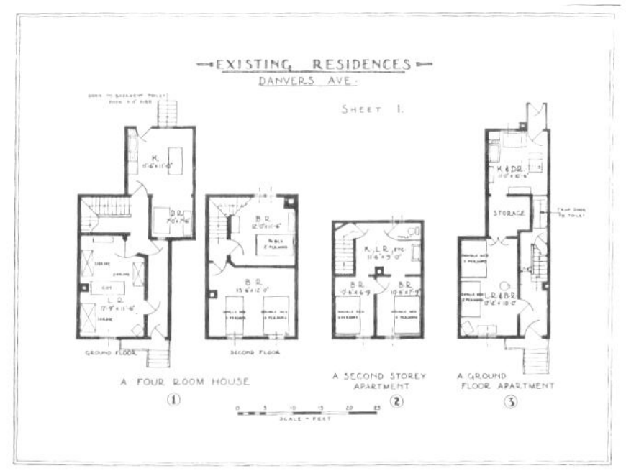
EXISTING RESIDENCES
DANVERS AVE.
SHEET 1.
K.
D.R.
L.R.
GROUND FLOOR
B.R.
B.R.
SECOND FLOOR
K. L.R. ETC.
B.R.
B.R.
K & D.R.
STORAGE
L.R. & B.R.
A FOUR ROOM HOUSE
1
A SECOND STOREY APARTMENT
2
A GROUND FLOOR APARTMENT
3
SCALE - FEET

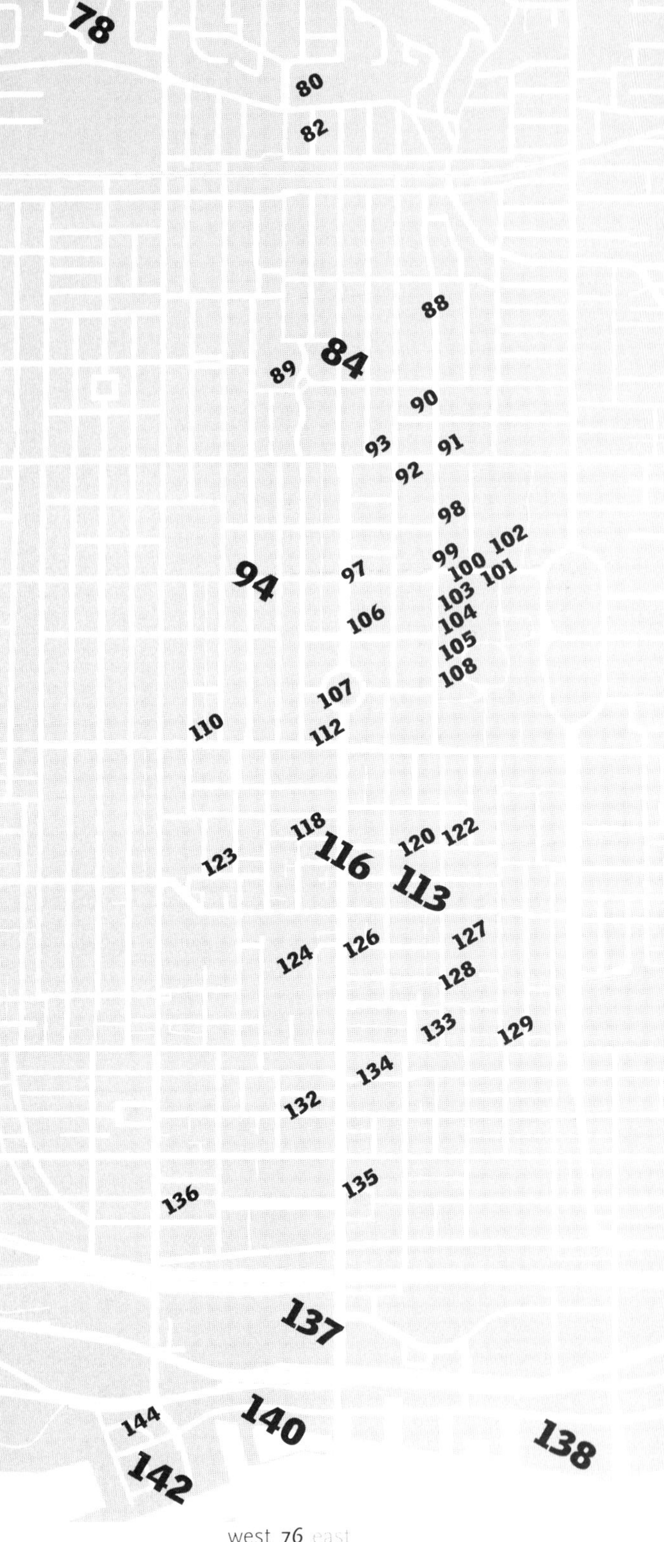
78
80
82
88
84
89
90
93
91
92
98
102
99
100
101
94
97
103
106
104
105
108
107
110
112
118
122
120
123
116
113
127
126
124
128
133
129
134
132
135
136
137
144
140
138
142

Wychwood Park

Wychwood Park sits on a height of land that was once the Lake Iroquois shore. The source for Taddle Creek lies to the north and provides the water for the pond found in the centre of the Park. Today, Taddle Creek continues under Davenport Road at the base of the escarpment and flows like an underground snake towards the Gooderham and Worts site and into Lake Ontario. Access to this little known natural area of Toronto is by two entrances one at the south, where a gate prevents though traffic, and the other entrance at the north end, off Tyrell Avenue, which provides the regular vehicular entrance and exit. A pedestrian entrance is found between 77 and 81 Alcina Avenue.

Wychwood Park was founded by Marmaduke Matthews and Alexander Jardine in the third quarter of the 19th century. In 1874, Matthews, a landscape painter, built the first house in the Park (6 Wychwood Park) which he named "Wychwood," after Wychwood Forest near his home in England. The second home in Wychwood Park, "Braemore," was built by Jardine a few years later (No. 22). When the Park was formally established in 1891, the deed provided building standards and restrictions on use. For instance, no commercial activities were permitted, there were to be no row houses, and houses must cost not less than $3,000.

By 1905, other artists were moving to the Park. Among the early occupants were the artist George A. Reid (Uplands Cottage at No. 81) and the architect Eden Smith (No. 5). Smith designed both 5 and 81, as well as a number of others, all in variations of the Arts and Crafts style promoted by C.F.A. Voysey and M.H. Baillie Scott in England. Between the two World Wars, a number of smaller houses were built when the Wychwood Park Trustees sold a portion of small lots along the western side of the Park. These houses varied stylistically from the earlier larger homes. After 1950, a few "modern" houses were erected on undeveloped lots.

In the 1980s the Park was threatened by the demolition of the large house at No. 16 for the purpose of redevelopment. This provided the impetus for the Park Trustees and other residents of the Park to seek designation of the Park as a Heritage Conservation District under Part V of the Ontario Heritage Act. After many meetings with Park residents and with the assistance of the Toronto Historical Board, a District Plan was approved by City Council and By-Law 421-85 was passed and approved by the Ontario Municipal Board in March 1986.

John Blumenson

MWF

Casa Loma

1 Austin Terrace
Architect, E.J. Lennox
Completed 1909–13

The gothic inspirations of Casa Loma

Built between 1911 and 1914, Casa Loma was home to financier and military officer Sir Henry Pellatt and his wife Mary. Through wise investments in electrical development, real estate and the Canadian Pacific Railway, Pellatt was one of the few men who were said to "own" Canada at the dawn of the 20th century. In 1905, he was knighted for his involvement in bringing electricity to the City of Toronto from Niagara Falls.

CTA. SC 244-4091

Pellatt engaged one of the foremost architects of the day, E.J. Lennox, to design Casa Loma and its associated buildings. Construction began on Pellatt Lodge (situated on the northwest corner of Walmer Road and Austin Terrace) and the Stables (to the north) in 1905. Upon completion, the Pellatts moved from their prestigious house on Sherbourne Street to the Lodge. From there, they were able to watch as construction on Casa Loma began.

The desire to build an ostentatious house was not uncommon in North

America- wealthy industrialists, such as the Hearsts and Vanderbilts, commissioned huge houses in the latter half of the 19th century. In Canada, however, Casa Loma was unlike anything ever seen before – it was the largest house ever built, comprising 180,000 square feet and costing Pellatt the princely sum of $3.5 million. By today's standards, Casa Loma would have cost Pellatt $44 million.

The architectural character of Casa Loma reflects the passion Pellatt held for the Gothic. The unusual combination of its elements (which include a three-storey bay window, a Norman and a Scottish tower, crenellations, heraldic beasts, Elizabethan-inspired plasterwork, and a 65-foot-high hammer-beamed Great Hall) draws from the Gothic and Romanesque styles. The effect created links Casa Loma to the first great Gothic Revival houses of 18th century England – Horace Walpole's Strawberry Hill and William Beckford's Fonthill Abbey, which demonstrate an imaginative interpretation of the Gothic. The fact that Pellatt had, in his painting collection, *A View of Fonthill Abbey* by J. M. Turner and *A Portrait of Sir Horace Walpole* by Sir Joshua Reynolds, further establishes the stylistic connection.

Despite the historic references, Casa Loma was fitted with the most modern conveniences of the early 20th century. Lennox shared Pellatt's interest in new and innovative technologies, such as electric lighting, heating and cooling systems, elevators, telephones and central vacuuming systems, and he had them all incorporated into the castle. Casa Loma was also embellished with exquisite plasterwork, beautiful wood flooring, and European marbles. No expense was spared on the materials used or the quality of workmanship.

The Pellatt's moved into their largely unfinished house in 1913, but the onset of the First World War halted construction. The economic downturn that followed the war further stalled the project and, with the collapse of the Home Bank of Canada in 1923, Pellatt's finances failed. Lady Pellatt died shortly thereafter. In 1924, unable to pay the municipal taxes on Casa Loma (which had risen from $600 per annum to over $1,000 per month in 1920), Pellatt was forced to suffer the heartbreak of selling off his personal belongings and collections at a five-day auction. He moved out in that year and, by the early 1930s, the City of Toronto took possession of the property. The interior of Casa Loma was never completed to Sir Henry Pellatt's original designs.

Joan E. Crosbie

Castle Hill Development

Completed 1991
Gabor + Popper Architects

Every corner of every city has a story to tell.

The history of Spadina Road between Dupont and Davenport tells a remarkable story of urban development, of social and political history, of poetry and personal tragedy.

The site of the Castle Hill development, south of the IROQUOIS Escarpment – the shoreline of a vestigial glacial lake that is now Lake Ontario, was occupied by more than one DAIRY: *Acme Farmer's*, and *Sealtest*. Just west, at the corner of Bathurst and Davenport, currently occupied by public transit service yards, commercial gardeners marked the landscape with FURROWS

● *Gabor + Popper*

to grow food for the local community. William Baldwin, who built and lived in Spadina House, was responsible for the SURVEY of Spadina AVENUE, Toronto's original grand avenue. Sir Henry Pellatt, Baldwin's neighbour at his eccentric Casa Loma, was president of the Toronto Power Company which provided both the physical and political POWER that contributed to the growth of the city. Documents that record these and many more local histories are housed in the Toronto ARCHIVES, located on the east side of Spadina, south of Davenport.

Every building in every city has a story to tell.

The history recalled by the Castle Hill development is one of a time and place far removed from the corner of Dupont and Davenport. Castle Hill's story evokes the Georgian period in England, stylistically interpreted by John Nash's neo-classical Park Crescent Terraces in London and by John Wood's Royal Crescent in Bath. Here, on Spadina Road, the Castle Hill marketing materials applied names such as "The Regency," "The Knightsbridge," and "The Cambridge" to the three-storey, stone-and-stucco row houses and spoke of a "Georgian-inspired community which echoes the old world overtones of Casa Loma ... every feature focused on making your life more comfortable and your entertaining more elegant."

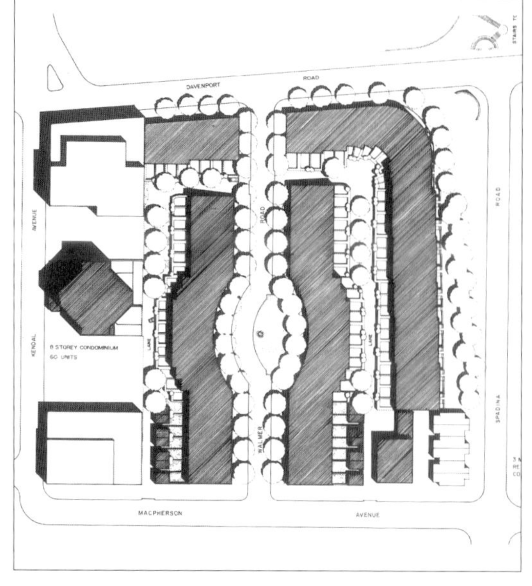

● Gabor + Popper

Because it attends to aspirations that are similar to Sir Henry Pellatt's Casa Loma, Castle Hill makes a questionable selection of the history it chooses to include and that which it omits. A few short years after the construction of Casa Loma, Sir Henry was unable to maintain the financial opulence of his dreams and was forced from his home. Rather than allowing the industrial history of the site inspire the project, Castle Hill reminds us of Sir Henry's tragic story of unsustainable folly.

Castle Hill has a story to tell.

One can only wonder why it is the recounting of the Georgian period in England and not the story of an evolving community and its relationship to the local landscape.

Words capitalized above are part of an artwork titled Spadina Line *(Brad Golden and Norman Richards, 1991) which recalls the history of the site.* Spadina Line *runs between Davenport and Dupont, on the west side of Spadina Road.*

Brad Golden and Lynne Eichenberg

The Annex

The Annex is one of Toronto's best-known and most-sought-after neighbourhoods – at least among those who prefer central-city living. Over 16,000 Annegonians occupy its approximately one-half square mile that runs from Avenue Road on the east to Bathurst Street on the west, and from Bloor Street north to the CPR railway lines. Simeon Janes, who developed most of the central Annex in 1886, referred to the area as the Toronto Annex, hence the unimaginative name. We can divide its past into three periods – up to 1913, to 1950, and since then.

MWF

Period 1: The formative era

The 1793 rural survey of York Township defined the future shape of the Annex. With Queen Street as the base line and Yonge Street as the north-south dividing line, Bloor northward to St Clair Avenue, one mile and a quarter, became the Second Concession from the lake. The first sideroad west from Yonge, also a mile and a quarter, became Bathurst Street. Within the Yonge-Bloor-Bathurst-St Clair block were five long, north-south, 200-acre farm lots, only a quarter-mile in width. The Annex covers much of the southern half of the block.

Within this rural template, beginning in 1857, speculators laid out their subdivisions. Most of the subdividing happened in the economically ebullient mid-1880s. In a few areas, subsequent subdivisions altered the original plans. Following the long, narrow farm lots, the subdividers created long north-south streets; however, not all east-west streets met one another, although most of the long, north-south blocks, are divided into three. The chief exception to the usual pattern is the strip east of Bedford Road, where the streets are oriented east-west, the result of the first subdivision. In 1883 that strip became part of the City of Toronto when the City annexed the Village of Yorkville. The City took over the remainder of the Annex in 1886 and 1887, providing it with public services such as water, fire protection, sewers, and pavements, the latter two paid in part through local improvement taxes. Building followed, but erratically and patchily.

EJR

As was the case everywhere in North American urban places, speculators laid out their plans, often well in advance of building. These "premature subdivisions" (as early 20th-century planners labeled them) usually preceded the buildings by one business cycle – in the Annex seventeen years, on average. Speculators sold piecemeal – to brokers who then sold to others, to individuals, and mainly to small-scale speculative builders who put up few houses at a time. The result was first a patchy pattern of houses, then in the next cycle, general infilling. In the Annex the process of building took over a third of a century – from the 1870s in the Yorkville strip to 1913 on the west side of the district.

Janes intended the central Annex, especially St George Street, to be upper income, as did the Baldwins, who initially laid out the curved stretch of Walmer Road. Both succeeded in persuading the rich to build villas, mostly in the late 1880s and into the early 1890s. Some of these successful businessmen also contributed large amounts for new churches in the neighbourhood, most of which still stand. Here and there on some other streets, upper-middle-class houses went up, but there were few lower-middle/working-class, bay-n-gable-style houses, so common south of Bloor. Janes and others also made it clear that commercial activity would be restricted to Bloor Street and other peripheral roads. After the depressed 1890s, building picked up again, but these later houses, while still substantial, were, owing to rising material and labour costs, largely a cut below the late-1880s style. A majority were semi-detached (which according to Patricia McHugh is the "Annex house"). Upper-middle-class Annegonians dominated the neighbourhood. Home ownership ran to 80 percent. Most tenants lived in the few low-rise apartments and industries were mostly located next to the railway tracks.

Period 2: Decline

From the 1920s to the 1950s, the social composition shifted. Although "invasion-succession" was much less pronounced than in, say, Chicago, larger houses gradually became rooming houses after the children of the affluent moved to Forest Hill and North Rosedale. Many households boarded University of Toronto students and single professionals and clerks. During the Great Depression and World War II, these trends intensified. Many homes were put to institutional uses, for example, as national head-offices such as the Ecumenical Institute on Madison Avenue. In the 1920s neither the first zoning controls nor the Annex Ratepayers Association (ARA) could stem the tide of change. By 1945, the population had risen to over 16,000 from 12,000 in 1923.

Period 3: Post-war rebuilding and stabilization

Other than some late 1930s infilling with mock-Tudor two-storey houses, little building had occurred since 1913. From the mid-1950s to the early 1970s, high-rise apartment structures replaced many of the elegant houses on large lots. The 1954 zoning system and the 1958 neighbourhood plan (the first in Toronto) had allowed the change. The Ontario Housing Corporation built the largest apartment block. Rooming houses continued to exist, and multiple occupancy of many houses lifted the population. The ARA revived to limit change and had its finest hour when the Province shelved the Spadina Expressway, which would have split the district down Spadina Road.

Since 1972, when a reform-minded City Council reduced development possibilities and the economy weakened, changes have been relatively minor. The city built three small social-housing projects. Conversions to single family residences gained ground, although many of the conversions house tenants. Many rooming houses became bachelorettes. Thirteen houses became group homes. Several condominium-tenure apartment blocks went up on the margins of the district. Parking has become more difficult and many residents have to buy street parking permits. An aura of stability seems to reign but, underneath, the residents are deeply concerned about the Province's reduction of public services and weakening of protection for the tenant majority.

Jim Lemon

MWF

CPD/CT

217, 228, 230, and 234 St George Street

Architect, George Popper

217 St George Street and the three historic houses across the street at 228, 230, and 234 St George typify the single family houses that once lined the grand streets of the Annex. The Annex, known as one of the City's first "suburban" neighbourhoods, retains many of its historic houses, typically now home to professors, writers, and university students. Recently, the historic houses at 217 and 230 have gone through a conversion to condominium housing units by Urban Corp.

The condominium units at 217 St George incorporate the historic house at the north end of the complex (note the original red sandstone columns at the entrance shared with Unit 14). The housing complex was designed to allow each unit owner a private street entrance. Owners of second-floor units also enter their unit at street level and immediately ascend a staircase to their two-storey unit with a roof top patio.

EJR

In the other condominium conversion project, at 230 St George, the developer links the three historic houses together with a new multi-housing unit to the rear to make one housing complex. Completed in 1996, the condominium complex retains the distinct appearance of the three original houses. The building at the south, 228 St George, was designed in 1901 by Arts and Crafts architect, Eden Smith (1858–1949). Originally from England, Smith established an architectural practice in Toronto and went on to design over 2,000 houses in the city, the best known of which are in the historic district of Wychwood Park. In the middle of the trio of houses stands a house designed in 1909 by local architects, Edwards and Saunders. At the north, 234 St George was constructed in 1903 as the home for Robert Watson, who commissioned one of Toronto's most influential architects, E.J. Lennox (1855–1933), for the design. Lennox was architect of the west wing of the provincial Legislative Assembly building at Queen's Park, the similarly styled Old City Hall (Queen and Bay streets), and Casa Loma (1910) at the top of Spadina Avenue.

Tamara Anson-Cartwright

44 Walmer Road

Completed c. 1965
Architect, Uno Prii

One of Toronto's most prolific designers of developer housing, Uno Prii was responsible for over 250 buildings, containing some 20,000 apartment units, built between 1957 and 1981. While many of these are indistinguishable from the general production of Modernist apartment towers, a few stand out for their distinctive sculptural approach and playful whimsy.

EJR

Several of these buildings are located in Toronto's Annex neighbourhood, interspersed among a dense fabric of 19th and early 20th century Victorian single family homes. Of these, 44 Walmer Road captures, perhaps better than most, Prii's playful flamboyance coupled with an otherwise conventional planning strategy, allowing for the construction of highly identifiable buildings within typical market driven budgets. The units are assembled to produce a rectilinear, cruciform plan point tower, which Prii adorns with a layer of curvilinear balconies sporting a circular railing motif. A playful canopy and fountain in the forecourt recall the exuberance of the Miami hotels of Morris Lapidus, bringing an unexpected lightness and joie-de-vivre to sometimes staid Toronto.

Although 44 Walmer and other nearby apartment towers by Prii were criticized for their typical Modernist insensitivity to the surrounding Victorian fabric, their idiosyncratic quality and lightheartedness, along with a renewed interest in Modernism, have earned them a certain cult status among young architects, and they continue to attract media attention.

Marco Polo

190 St George Street

Architect, Joseph A. Medwecki
Completed 1972

One of the most arresting images of the early Modern period has to be the Mies van der Rohe photomontage of curvilinear towers of stacked concrete floors completely enclosed in glass. The purity of this image tantalized generations of architects, and indeed the problem of realizing the idea of this kind of transparency was central to Mies's entire career in building. The problem was that one could never achieve that kind of actual transparency. The architect had to use other means to achieve the ideal. (There have been architectural problems like this since the Egyptians tried to build reed houses out of stone.)

But with transparency, the lure of literalness is always present, that is, confusing the use of glass as equaling the effect of transparency. Also, the inevitable encumbrances of window frames, the need for privacy, the backs of refrigerators, etc., always get in the way. It has only ever been achieved at great cost and under high-art conditions by someone like Mies who really knew what he was doing. Only achieved by Mies that is, until the great, late-Modern trick of the continuous balcony was discovered. (And I admit here that I don't know who came up with the trick first ...)

EJR

By pushing the floor slab edge and the outer column line beyond the tower's facade, the eye is refocused on the structural patterns and away from the enclosing walls, which are then free to accommodate the contingencies and impurities of human occupation. The squinting eye can almost recapture the giddy energy of the early 1920s when people were declaring that "new living demanded new forms."

190 St George is an elegant solution to the continuous-balcony apartment type. The bold, white slab edges and exposed structure stand out crisply against the simple steel handrails and mostly glass enclosure wall – but I understand the apartments are difficult to furnish or hang pictures in because they're mostly glass-walled. The prow of the east and west balconies add a suitably minimalist-expressionist touch.

The building is one of the earliest condominium developments in the city, which may account for its better level of finish and well-kept appearance (condominium corporations having to take care of their investment by law).

It is also a reminder of a time when a new development strategy could link itself with the notion of progressive architecture – and still be considered marketable. 190 St George shows up the more recent Georgian-style condo-hulks being erected in the name of "preserving resale value."

[Also notable is the pair of continuous-balcony gems at 10 Avoca Avenue, southeast of Yonge and St Clair (1971, Seligman and Dick Architects).]

Ian Panabaker

George Gooderham House (York Club)

135 St George Street
Architect, David Roberts Jr
Completed 1892

The house at 135 St George Street was completed in 1892 for George Gooderham (1820–1905). Gooderham was the president of the Gooderham and Worts Distillery, a family-owned company founded in 1832 and, by the late 19th century, the largest enterprise of its kind in the British Empire. In addition to his role in industry, Gooderham served as president of the Bank of Toronto and of several insurance companies. The Toronto General Hospital, the University of Toronto, and the Toronto College of Music were among the institutions to receive his philanthropic support. Gooderham financed many of Toronto's landmark buildings, including the King Edward Hotel (1902) and the famous "Flat Iron Building" (1892), the headquarters of his business empire. He commissioned Toronto architect David Roberts Jr to design his residence on Bloor Street West, at the south end of the popular Annex neighbourhood.

EJR

The George Gooderham House displays the hallmarks of Romanesque Revival design, the most popular style of the time, with castle-like forms, round-arched openings, and elaborate sandstone detailing. The sprawling plan is anchored by a corner tower and, off St George Street, a porte-cochere. Toronto sculptors Holbrook and Mollington created the intricate carvings, with grotesques and human faces (including one of architect Henry Sproatt, who collaborated on the design). The elaborate interiors incorporate finely detailed wood finishes, a monumental three-storey staircase, and parquet floors with mahogany inlay. Gooderham named the house "Waveney," after a river near his birthplace in Norfolk, England. Gooderham resided here until his death in 1905, when he was described as the wealthiest man in Ontario, with a personal fortune of $25 million. The York Club, a private club for gentlemen, has owned the property since 1910. The George Gooderham House is the last surviving example of the mansions that lined Bloor Street in the late 19th century.

Kathryn Anderson

Rochdale College

(now the Senator David A. Croll Apartments)
341 Bloor Street
Architects, Tampold and Wells
Completed 1968

Rochdale College, together with nearby Tartu College, brought a characteristic 1960s-style, high-rise dormitory architecture to the northern fringe of the University of Toronto campus – a genre introduced by Tampold and Wells in their earlier Student Family Housing towers at 30 and 35 Charles Street West – but, in Rochdale's case, history and legend far outstrip anything that architecture might provide. Conceived as a combination of student co-op and "free university" experiment, it quickly descended into an unruly, anarchic state and became notorious as *the* nexus for Toronto's counterculture. After innumerable police raids and drug casualties, what remained of Rochdale was closed down in 1975; following extensive renovations, the building was matter-of-factly reborn as a seniors' apartments in 1979, but the rosy-lensed heart of "the runaway college" in all its colourful legend has, and probably for the better, not died so easily (as evidenced by its numerous subsequent histories and retrospectives, as well as the happily vestigial "Unknown Student" sculpture up front).

It seems paradoxical, in retrospect, that an icon of "anti-establishment" culture was housed in a building that, within the context of 1970s urban reform, must have epitomized bad old "establishment" ways in architecture and urbanism. (For a useful contrast, refer to the adjacent Sussex-Ulster neighbourhood, much of which was reincarnated as student co-op housing in the wake of the urban reform movement.) One can sense that the apparently frank concrete Brutalism of Rochdale highlighted the offbeat harshness as well as the creative fervour of the activities within. Yet, in its current happy afterlife as the Croll Apartments, Rochdale presents a surprisingly urbane aspect. All things considered, its Brutalism is fairly restrained, the offset tower mass (not unlike that of Tampold and Wells's earlier Charles Street apartments) is attractively proportioned, and the corner plaza can be seen as a positive contribution to the Bloor streetscape. Even the retail-related alterations and fine-tunings over the years don't seem to have compromised Rochdale's fundamental lines. We may not build cities or universities like this anymore, but a third of a century after its conception, this oft-mythologized landmark deserves, in its own right, a certain appreciative respect.

Adam Sobolak

EJR

Tartu College

310 Bloor Street West
Architects, Tampold and Wells
Completed 1969

In scale (18 floors), style (Brutalist), and function, Tartu College is a two-year-younger sister building to Rochdale College. It was built as a general undergraduate student co-op combined with a library, archive, and study centre serving the Estonian-Canadian community. Having escaped the melodrama that befell Rochdale, Tartu remains to some extent a microcosm of dormitory fashion of its time. The basic housing unit is a spartan six-person suite, and there are five suites on each residential floor. The arrangement is fundamentally modular, as was fashionable in student housing during the 1960s and 1970s. Surprisingly little of this type of student housing was built in the University of Toronto environs (due to funding cutbacks and changes in the political and planning climate), although it is a leitmotif of newer campuses, such as York, Waterloo, and Trent universities. Tartu's compact, conventional apartment-building form overcomes the overwrought, Skinner-maze effect that plagues many of its contemporaries to a greater or lesser degree. Meanwhile, the Corbusian architectural overtones remind us of how Le Corbusier's "machine for living" ethic achieved some of its purest, most appropriate expression in 1960s and 1970s student housing – as those who've spent any time in it will readily attest.

EJR

A somewhat purer, if less self-consciously monumental rendition of concrete Brutalism than Rochdale (with Aaltoesque "Scandinavian" traits that must have suited the building's sponsors), Tartu also suffers more from the heavy-handed urban narcissism that soon gave Brutalism a bad name; its entrance is overly dark, steep and tight, and the building turns its back to the corner of Bloor and Madison, a far cry from the generous corner plaza offered by Rochdale. On the other hand, these sins are easy to overlook because Tartu has survived the years and can now be cherished for the nature of much of its detailing – most especially, the superscaled, cast-in-place concrete lettering flanking the entrance. Ironically, Tartu's placid history has led to its becoming, in lieu of Rochdale's ghost, an evocative reminder of the long-gone, sideburns/bell-bottom/Wallabee era in university culture.

Adam Sobolak

Sussex-Ulster Residents' Association

One day in 1968, the residents of Robert Street, south of Bloor, woke up to find that a whole block of houses was being torn down. The zoning called for high-rise, like St James Town. Two 20-storey buildings were erected, but the residents fought another two to a draw. Eventually, the developer gave up and traded for the University of Toronto Aura Lee Field beside Ramsden Park. The University got a field almost on campus and the developer got two towers overlooking a park.

From that feisty beginning, the Sussex-Ulster Residents' Association has taken on some big battles and scores of little ones. When issues arose, the residents of 1,300 homes, from Bloor south to College and between Spadina and Bathurst, have always had a committee to go to City Council or the Ontario Municipal Board. Early in the 1970s for example, we got a traffic maze even though the Works Department wanted faster streets. Council voted for one-way streets with a turn at every corner, which brought about a 30% decrease in cars and lower speeds that have reduced accidents dramatically. Despite ten years of successful experience with the northern maze, however, the residents south of Harbord had to fight for their maze for five years.

In the early 1980s we prevented a bank from tearing down a terra cotta gem at Bathurst and College. In 1987, we celebrated the City's 150th anniversary by getting plaques for many of the fine century townhouses. We sent around a pamphlet explaining that "your house is worth $10,000 more if it has the original facade." The neighbourhood has seen most of the older immigrant groups migrate to bigger lots in the suburbs, and new, younger families arrive and convert the student rooming houses to single-family homes. They strip the brick and rebuild the porch detail. At least two have rebuilt slate roofs. What were primary colours have retreated to subtle Victorian shades.

Perhaps our most dramatic battle was against Doctors' Hospital. We fought for ten years, including three and a half weeks at the Ontario Municipal Board, against a half-block, 87-foot-high complex facing low-rise houses on both sides. The fight cost $12,000. We lost, but by the time the Doctors' Hospital "won," the provincial government had changed, hospitals were being amalgamated, and the project was cancelled. Now, another ten years later, they want to create "extended" care beds for "nursing home" care.

Most of our fights have been on the Spadina side where University proposals continue to intrude into the residential neighbourhood. There have been several plans to build in the playground on Robert Street. None have been built so far. It would be ironic to fight off a developer's highrise, but get another from a humanistic institution such as the University. Just last year the University erected "the giant O" student residence. They brought in a famous American architect to erect what most old-time residents see as an ugly blot on the street. A herd of architects arrived at the public meeting to try to convince us that jarring is beautiful. What with "Fort Book" (the Robarts Library) and "Fort Jock" (the Athletic Centre) along Harbord, the lessons of history are slow to be learned.

Bloor Street has become what one might call a restaurant anthill. The bad old days saw us fighting off screaming music from open windows and wet-tee-shirt contests at local pubs. Who hasn't had their fence kicked in, listened to fights or cleaned up after party-goers? The Liquor Licence Board may have turned down an application for another dance hall, but we now have a photocopy centre open twenty-four hours. We beat off the bingo parlour, closed the rooftop patio, but succumbed to the betting shop. Nightclubs have brought generations of "newest look" teenagers. Cafes are found on every corner. After we reached 52 restaurants, the residents demanded and got a new by-law restricting restaurant size and ensuring that parking be built. Homeless folks have now decided that Bloor Street is a great place for getting change from wealthy diners.

EJR

The story of our neighbourhood would be incomplete without the tales of driving round and round looking for a parking spot. You give up and park on the wrong side. You get a ticket, just like the visitors who have taken not only the legal spots, but parking all along the other side. It is virtually impossible to get a fire truck through.

The homeless shelter, temporarily at Doctors' Hospital, took a hundred people off the streets in the winter of 1998. Some folks thought the neighbourhood was doomed. Surprise ... we all got along. As we did when House Link built a low-rise apartment for former psychiatric patients. No problem.

We were told the neighbourhood was a slum in 1973. We couldn't get a mortgage because "the houses were all run down and made out of wood." We were told the area would be redeveloped, just like St James Town, and we'd make a fortune. But the mission of the Sussex-Ulster Residents' Association was to restore to life our part of the inner city. And we have!

Bob Barnett

Campus Co-op

Campus Co-op has over 32 houses in the Annex neighbourhood. The main office is located in the Arthur Dayfoot House at 395 Huron Street.

In the dirty thirties, life was harsh in Canada. It was a time of extreme poverty for the average person. Hardly the time to be a student when no loan or grant programs existed, let alone to start a radical new idea in student housing: co-operative living. The four founding members, Arthur Dayfoot, Archie Manson, Donald McLean and Alex Sim, were young students committed to the Social Christian Movement (SCM). From libertarian theologians, such as Toyohiko Kagawa and Fathers James "Jimmy" Tompkins and Moses Coady, these students learned about the Rochdale Principles of co-operation, then applied them in the creation of the longest running housing co-op in Canada: Campus Co-op.

Campus Co-op was started in 1936, two years before the internationally famous Tompkinsville in Cape Breton, Nova Scotia, completed its first house. Since its inception, the co-op has been an important vehicle for co-op housing expansions, helping to found over 100 co-ops in Canada, Africa, and the United States, including Toronto's Rochdale College in 1968.

EJR

What makes this story compelling are the people who were involved with Campus Co-op's various stages. The students founding of the co-op, in a time of economic adversity is remarkable. Involvement in the explosive growth of the co-op movement in the 1960s was a formative experience for many of that generation's political and artistic leaders; Ed Broadbent is a notable example. People who first learned about co-operative housing from Campus Co-op living, helped start and run the co-op housing sector in Canada from the 1970s to the 1990s.

Joey Schwartz

Graduate House

633 Spadina Avenue
Architects, Morphosis/
Stephen Teeple Architects
Landscape Architect, Janet Rosenberg
Completed 2000

Graduate House is Toronto's first aggressively deconstructivist building. Architect Thom Mayne, from the Los Angeles firm Morphosis, collaborated with Toronto-based Stephen Teeple to win an invited design competition for the project in 1998. Strict site guidelines and a rigid volumetric envelope were prescribed by the City of Toronto, generating major design challenges; and the University of Toronto required that the project be realized quickly on a modest budget.

EJR

The architects responded with a set of tightly packed, efficient urban blocks composed around a central courtyard and reflecting pool. Graduate House accommodates 450 students in three- and four-bedroom apartments and includes a restaurant at the southwest corner. The building is clad in charcoal-coloured precast concrete with a perforated aluminum skin draped over the north and east facades. A skew in the south block generates positive agitation and "delaminates" the facade into overlapping planes of texture. The signature feature of Graduate House is a multi-valent, two-storey truss element with the words "UNIVERSITY OF TORONTO" featured in Pop-scale, fritted glass lettering. This dramatic, tectonic element cantilevers over Harbord Street and serves simultaneously as cornice, corridor, lounge, sign, and western gateway to the University of Toronto's St George campus. This controversial project has won three design awards from: the *Canadian Architect*, *Progressive Architecture*, and the Los Angeles Chapter of the American Institute of Architects. Architecture critic Christopher Hume called the Graduate House Toronto's "first architectural landmark of the 21st century."

Larry Wayne Richards

Innis College Residence

111 St George Street
Architects, Zeidler Roberts Partnership
Completed 1994

Completed in 1994, this new complex is the first permanent home for students enrolled at Innis College, situated directly opposite on the west side of St George Street. The form and silhouette of the residence was generated by what the architect Eberhard Zeidler has described as "a new fractal geometry." The predominant factor that determined the forms was the existing array of historic Victorian and Edwardian houses that line the street to the north and to the south. To reduce the apparent scale of this six-storey complex and break down the bulky form, the architects designed the facade as four seemingly separate buildings whose massing is more sympathetic to the smaller domestic scale of the adjacent houses. Familiar Toronto building materials such as reddish-brown brick, light grey concrete, and grey roofing materials were used in a convincing manner to allude to the domestic communal "house" form of nearby student accommodation such as Devonshire House, Whitney Hall, and the Sir Daniel Wilson Residence. The architects purposely sought to introduce a complexity of architectural elements (as they had done in the adjacent Rotman School of Management to the south completed in 1995) that combine a variety of window expressions and a tripartite stacking of materials (stone base, brick body, and gable).

Robert G. Hill

The project contains 46 four-bedroom, 28 five-bedroom, and 8 two-bedroom apartment-style suites arranged in a U-shaped configuration around an outdoor court. Interior corridors were arranged so that students would pass through common areas or "living rooms" that are visible from the street and the court. The common areas create a sense of home and community within each grouping of rooms.

Robert G. Hill

W.D. Matthews House (Newman Centre)

89 St George Street
Completed 1890
Alterations 1899, George M. Miller

On the northeast corner of Hoskin Avenue and St George Street there is a handsome and richly clad house that every University of Toronto student should know. It's a red brick, three-storey modeled in the style of the Romanesque and its basement houses a coffee shop that is a calculated distance from Robarts Library across the street.

EJR

The residence was built in 1890 for "The Barley King of Canada," Wilmot Deloui Matthews. Matthews was the co-founder of the Canada Malting Company and President of Consolidated Mining, the Canada Foundry, and the Pembroke Railway. His residence is one of the many large estates that lined St George Street at the turn of the century.

To celebrate the occasion of his daughter's wedding, Matthews commissioned Toronto architect George M. Miller and sculptor Gustav Hahn to design an Art Nouveau ballroom for his house. Following his death in 1919, the house and its ballroom became a residence for Catholic students; in more recent years, it has been occupied by the Newman Centre Chapel.

Despite the changes in ownership, the Matthews House still retains its original exterior features and, most remarkably, the splendours of its interior. Whereas the future is uncertain for many historic homes in Toronto (such as the residences on Jarvis Street and Wellesley Place), stewardship of this residence illustrates that re-use of our historic built environment can, in fact, successfully preserve it.

Nancy Byrtus, Judy Matthews

Massey College

4 Devonshire Place
Architect, Ron Thom
for Thompson Berwick Pratt
Completed 1963

Perhaps in response to the legendary reputation of nearby Devonshire House residents, Ron Thom's well-documented and adored Massey College happily banishes the outside world and creates a handsome,

EJR

small-scaled, cloistered one within its walls. Prismatic and planar on the outside, more open, fluid planes form the courtyard inside. The intent, in the words of the founder Vincent Massey, was to provide a home "for a community of scholars whose life will have intimacy but at the same time academic dignity."

David Winterton

Devonshire House

3 Devonshire Place
Architects, Eden Smith and Son
Completed 1908–09

Originally built to alleviate a student housing crisis on campus, Devonshire House (begun in 1906) was completed just in time to become a residence for the Royal Flying Corps during the Great War. Three sombre sandstone buildings form an open-ended quadrangle. A fourth residence, fronting Devonshire Place, was to be built "later." When Devonshire House evolved into a residence for professional-faculty men students, the lack of the fourth building provided an open stage for nine decades of collegiate antics, ranging from

EJR

avant garde engineering experiments and nude calisthenics, to the endless whoosh of water bombs from third floor windows. This year, "Devo's" transformation into the (non-residential) Munk Centre for International Studies (Kuwabara Payne McKenna Blumberg Architects) should be complete.

David Winterton

Trinity College

6 Hoskin Avenue
Architects: Darling and Pearson (1923–25);
George and Moorhouse (1940–41);
Sir Giles G. Scott (1953–55);
Somerville McMurrich and Oxley (1963)

Founded in 1851 by Bishop John Strachan and formerly located on Queen Street West, the "new" Trinity College on Hoskin Avenue is actually the second site for the complex. This Church of England College offers undergraduate programmes in the arts in affiliation with the University of Toronto and grants its own degree in theology. Designed by the renowned Frank Darling (1850–1923), the ambitious Master Plan for the new site was presented in 1915 but construction of the first phase was delayed nearly eight years and not completed until 1925, two years after the death of the architect. The plan of the new ensemble mimics (but does not replicate) the old complex, and new materials such as Credit Valley stone (instead of brick) and concrete floors and stone staircases (instead of wood) were effectively employed throughout the building.

Robert G. Hill

Nearly fifteen years later, the Toronto firm of Allan George (1873–1961) and his partner Walter N. Moorhouse (1884–1977), designed impressive additions to the complex including the West Wing, with the Strachan Dining Hall, and the East Residence Wing containing Whitaker House, Welch House, Body House and the striking Henderson Tower which features a spacious Tudor portal leading to Philosopher's Walk.

In 1953–55 the eminent London architect Sir Giles Gilbert Scott (grandson of Sir George G. Scott) designed the exquisite chapel now fronting on Hoskin Avenue. The north side of the quadrangle was finally enclosed and completed with the construction of Cosgrave House in 1963, designed by Somerville, McMurrich and Oxley. The entire ensemble thus became the closest evocation one may find in the city to the scale and atmosphere of the classic college quadrangles found in Oxford and Cambridge.

Robert G. Hill

Whitney Hall Residence

85 St George Street at Hoskin Avenue
Architects, Mathers and Haldenby
with consulting architect John M. Lyle
Completed 1930–31

In 1930 the western boundary of the University of Toronto campus was formed by St George Street. To establish a consistent style and character to the public face of the institution, the administration mandated that all new buildings along College Street and St George Street should be "Georgian in character." When the commission for Whitney Hall, the new residence for women students of University College, was awarded to Mathers and Haldenby, the firm responded with an appropriately conservative plan and an impeccably detailed elevation (*R.A.I.C. Journal*, ix, May 1932, 114-20, illus.). The same architects would continue this tradition nearly 25 years later with their design for the Sir Daniel Wilson Residence (1953–54).

Whitney Hall is comprised of three "houses" each accommodating 50 women in single and double rooms. The first three units, Cody House, Falconer House, and Mulock House, were completed in 1931 but the fourth unit, Ferguson House, which replicated the style of the original neo-Georgian scheme, was not built until 1960 and added to the sense of enclosure of the quadrangle. A unique aspect of the plan is the inverted nature of pedestrian access – entrances to the various houses were located within the courtyard, rather than on the street side, and special "night doors" leading to the basement of each house were provided for students returning late at night, but only after they had obtained special permission from the House Don!

The residence honours the name of lumber baron Edward C. Whitney, brother of Ontario Premier Sir James P. Whitney, and was built with a bequest from his estate. All the furnishings were purchased using funds raised by the graduate and undergraduate women students of University College.

Robert G. Hill

Robert G. Hill

Sir Daniel Wilson Residence

73 St George Street
Architects, Mathers and Haldenby
Completed 1953–54

Opened on December 4th, 1954, the $2 million University College Men's Residence honours the name of Sir Daniel Wilson (1816–92), who served as principal of University College from 1880 and later became President of the University of Toronto from 1887 to 1892. It was one of the last structures erected on St George Street in a genteel Georgian Revival style prescribed by the University after 1920 which was followed for more than thirty years, until the campus building boom on the west side of the street after 1955. Designed by the firm of Mathers and Haldenby (who also planned Whitney Hall to the north), the project has 183 single rooms and a dining hall seating 220 persons (*Globe & Mail*, 30 Nov. 1954, p. 3, illus. & descrip.). The plan was divided into six houses named after the heads of the College from the time of its establishment in 1853 (McCaul House, Loudon House, Hutton House, Wallace House, Taylor House, and Jeanneret House), and each offered the requisite comfort and convenience (but not extravagance) one might expect in post-war campus buildings.

Robert G. Hill

The U-shaped plan was an attempt to form a quadrangle with the existing University College (1856–59) on the east side, but the degree of enclosure was no match for the comforting sense of containment and scale found in the quadrangle of nearby Trinity College. The project redeemed itself, however, by its distinctive clock tower facing St George Street, which has been set above the stripped-down classical portico entrance reminiscent of the work of Robert Adam and George Dance, two eminent architects of Georgian London.

Robert G. Hill

Macdonald-Mowat House

63 St George Street
Builder, Nathaniel Dickey
Completed 1872

At first glance this residence may escape notice due to the prominence of the buildings that surround it. Robarts Library and the School

EJR

of Graduate Studies draw your eye to the north and Knox College lies to the south. Constructed in 1872 by the owner and iron founder Nathaniel Dickey, the design of this residence hybridizes elements of Italianate and Second Empire styles. Due to its south-facing entrance, it presents a somewhat blank face to St George Street despite the projecting eaves and articulated keystones that grace its west elevation. The history of the Macdonald-Mowat House is what really sets it apart, as an Ontario Heritage Foundation plaque explains:

> Sir John A. Macdonald, Canada's first prime minister, purchased this house in 1876 and lived here [from] 1876–78 … Macdonald owned the property until 1886 and it was occupied by his son, Hugh John, [from] 1879–82. The Hon. Oliver Mowat, prime minister of Ontario, bought and occupied the house in 1888 and retained ownership until 1902. The property was leased [between], 1897–98, to the Hon. Arthur Sturgis Hardy who succeeded Mowat as prime minister, and sold to Knox College in 1910.

Nancy Byrtus

Plaque text courtesy of the Ontario Heritage Foundation

New College

300 Huron Street
Architects, Fairfield and DuBois
Completed 1964, 1967

New College, designed by Macy Dubois and built in two phases during the 1960s, was and remains one of the finest buildings on the University of Toronto campus. Within the housing field, it is, like Massey College, a landmark building, which broke away from the then current principles of modernism by respecting the context of the street and by providing enclosed outdoor spaces.

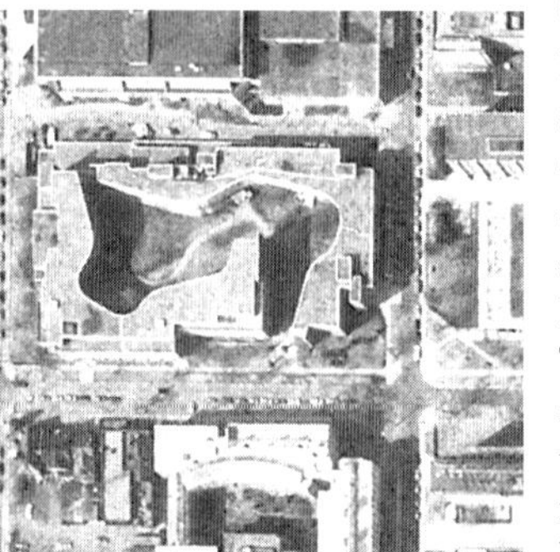

Metropolitan Toronto, 1992 (now City of Toronto)

New College is graced by well-detailed, curvilinear facades built in a warm-coloured brick that has an almost sensual materiality. New College's design is indebted to the Massachusetts Institute of Technology's Baker House and Aalto in general, especially in the detailing of the interior spaces, which have large, natural-wood profiles and exposed brick (the same as on the exterior). The building has aged exceptionally well and is still in very good repair, with few changes made to the original design.

The only negative aspect of the College is a curious muteness towards Spadina Avenue: windowless walls on Spadina and Classic avenues appear rejecting and dull. This was by design, because at the time of planning, the proposed Spadina Expressway threatened to destroy the residential character of the neighbourhood, and the architect wanted to turn the active part of the building away from noise and pollution.

The Spadina Expressway was never built and, in recent years, Spadina has been transformed into an attractive tree-lined boulevard. Fortuitously, New College is presently looking at possibilities to enlarge its residential capacity. After forty years, the College may be able to reconsider its relationship to Spadina Avenue. A study conducted last year by the Faculty of Architecture, Landscape, and Design, with Macy Dubois's concurrence, concluded that it is feasible to add rooms to the west and north sides, thereby opening up the blank walls and giving the College a much better civic face to the boulevard.

● *Klaus Dunker*

Klaus Dunker

Knox College, Spadina

(also known as the former Connaught Laboratories)
1 Spadina Crescent
Completed 1873
Architects, Smith and Gemmell

Founded in 1844, and renamed Knox College in 1846, this Presbyterian theological institute retained architects Smith and Gemmell in 1873 to design a Gothic Revival structure in the centre of Spadina Crescent. Completed two years later, the distinguished High Victorian edifice became the home for the College for the next 31 years. The building boasted a convocation hall, classrooms, a library, and dormitories for almost 100 divinity students. With its landmark location, bermed landscape, and southern orientation the former college building attracts attention away from its more modern surroundings. Knox College must have thoroughly dominated the 19th century urban setting in the days of horse and buggy.

The building is composed of an axially symmetrical centre block and square tower, with supporting gabled wings to the east and west. Among the medieval-inspired elements are the pointed windows, angled buttresses on the walls, rounded buttresses on the towers, engaged columns, finials, gargoyles, corbel tables, and steeply pitched dormers. The buff brick walls are divided by a series of horizontal drip courses and window hoods of sandstone. The windows, grouped in larger brick arches, diminish in size from the ground to third storey, increasing the sense of height. The circular motifs centred on the upper third of the lancet window mullions add interest and bring unity to the fenestration. Much of the interior has been altered, but the groin-vaulted main vestibule at the south, arched halls on the ground floor, and central oak staircase remain as faint echoes of the Gothic exterior. The overall architecture is restrained, conservative, and markedly institutional.

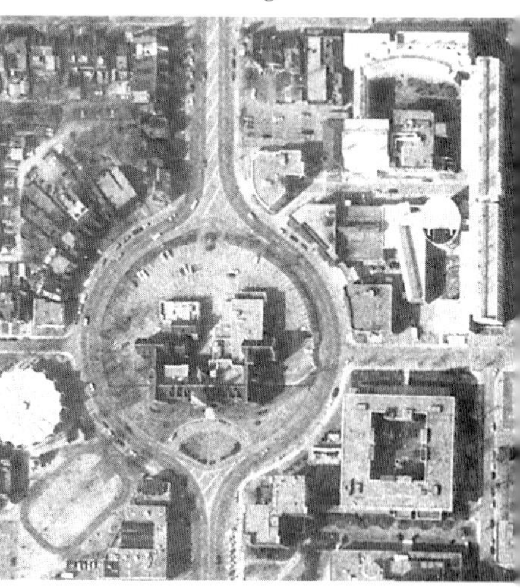

Metropolitan Toronto, 1992 (now City of Toronto)

The College building later became the Spadina Military Hospital, then, after being repurchased by the University of Toronto in 1943, the Connaught Laboratories – famous for its historical association with the development and production of penicillin and insulin. In subsequent years there have been several additions to the rear of the main block, most of which have been less than sympathetic to the original design. Today the building is occupied by various university departments, medical laboratories, and other campus organizations.

Sean C. Fraser

Knox College, St. George

59-61 St George Street/
23 King's College Circle
Architects, Chapman and McGiffin
Completed 1915

In 1887, as a result of the *Federation Act*, Knox College was incorporated into the University of Toronto. In 1906 the College moved out of 1 Spadina Crescent and into buildings on St George Street. Planning soon began on yet another facility, this time located between King's College Circle and St George Street (59–61 St George Street and 23 King's College Circle). Constructed between 1912 and 1915 of grey sandstone with dressed limestone trim, this Perpendicular Gothic-style collegiate building, by Chapman and McGiffin Architects, is an excellent example of a medieval-revival academic residence and the design skill of Alfred Chapman.

The bold plan features a small grassy quadrangle with a central cloister joining the academic and spiritual areas with the residences. The walls within the quad, and on the exterior, are interrupted by a series of tall bay windows that spring from simple buttresses at the second storey. Hipped dormers, clad entirely in green slate to match the gabled roofs, reveal the residential nature of the south and west wings. Approximately 100 men and women are accommodated throughout the 3 1/2 storeys. A tower, at the centre of the west facade and the tallest part of the College complex, identifies this important academic and ecclesiastical landmark on the University of Toronto campus. With its robust buttresses, leaded glass windows, and powerful massing, the College provides a strong and architecturally interesting western edge to King's College Circle.

The central rotunda, located in the middle of the eastern wing, has a vaulted ceiling, fan leaf tracery, octagonal piers, and stone steps flanked by richly carved and solidly built stone rails. To the south, pink marble stairs lead to the jewel of the College – the Chapel – with its vaulted Gothic ceiling, stone piers, fitments, and great window. On the north side of the rotunda, at the top of another set of stairs, is the academic heart of the college, the 70,000-volume Rev. William Caven Library, noteworthy for its oak ceiling supported by a series of elaborate, wood hammer beams. The library's namesake was the principal of Knox College (1873-1904), the driving force behind the building of 1 Spadina Crescent (1873-5) and the president of the Pan-Presbyterian Alliance (1900-1904).

EJR

The College is richly appointed with English Gothic ornament. On the east and west elevations comical mascaron grace the upper reaches of the architecture. Large, arched metal doors, completely glazed with small divided panes and maneuvered by ring knockers, also add to the medievalism of the place. Like the ancient English universities from which the architecture derives, at Knox College religion, academia, and a monastic sense of serenity combine in what is arguably Toronto's most architecturally traditional university residence.

Sean C. Fraser

Peregrine Housing Co-operative

18 Grenville Street
Quadrangle Architects Limited
Completed 1996

• *Quadrangle Architects*

This co-operative housing development was the first to be developed under the City's East of Bay Part II Plan, which sets down specific building envelope guidelines. These included the creation of a strong cornice line at the fourth floor, and steps at the eleventh and fifteenth storeys. The building contains 174 suites and three levels of underground parking, and it incorporates the new headquarters of the Ontario Civil Service Credit Union, which had previously been located on the site. Despite the tight site, the building offers outdoor amenity spaces, available to all residents, on the fourth and fifteenth floors.

Leslie M. Klein

Live/work loft conversion on Croft Street

16 Croft Street

Originally, Croft Street was a significant north/south thoroughfare connecting Bloor Street to College. As the area around it became more developed, it took on many of the functions typical of Toronto lanes, such as the provision of garages and vehicular access to the houses flanking Croft east and west; however, a number of houses, coach houses, and warehouses remain that reflect the street's previous primacy.

MWF

No. 16 Croft Street is part of a larger building that was converted into five freehold residences. Prior to the renovation, it housed a Turkish rug cleaning business that had been in operation since the 1920s. Before that, the building was a munitions factory during the First World War. A group of people (all the end users), purchased the warehouse in 1987 with the intention of converting it to live/work spaces. The existing building had 10,000 square feet of open space, on two floors, with windows on all four sides. The masonry walls are a mix of clay brick on the two exterior wythes and concrete brick on the third, interior course. The structural system is a combination of timber and steel beams, and mill-flooring decking throughout. There is no at-grade outside space on the property, as the property lines mirror the floor plate of the building.

The proposed conversion from existing non-conforming use to residential was turned down at the Committee of Adjustment in the fall of 1987. It was appealed and approved at the Ontario Municipal Board in the spring of 1988.

The building was divided vertically to create five equal houses, each with a total area of about 2,000 square feet. Three large arches were cut out of the front masonry wall on Croft Street and five parking spots were carved into the previously interior ground floor area. The newly created exterior space also provided a recessed entry area and storage facilities for each house. The square footage removed from the ground floor to provide access and parking was "relocated" as a continuous strip along the mid-third of the existing roof. The new party walls defining each house were continued up through this volume, making a new third-floor room flanked on either side by decks. The siting and massing of the third-floor volume was designed to minimize the impact on the neighbouring houses, both in respect to privacy and shadows.

The ground floor of 16 Croft Street was designed to work as an independent office or apartment, with its own entrance. A double-height space at the back half of the space was created to allow for a small sleeping loft above the washroom/laundry/storage area. In addition, there is a large foyer and stair, separated from the ground-floor apartment, that serves as the entrance to the second unit above.

The second floor is organized into two distinct areas defined by the placement of stairs and other "objects" that create various degrees of enclosure. Changes in the floor levels/ceiling heights mark the transition between the two areas. The public precinct accommodates the dining room, living room, kitchen, and small sitting area, and has 14-foot ceilings. The private zone is up three steps and contains the bathroom and two bedrooms. The ceilings in this area are eight feet high, with the windows set low to the floor. A staircase leading up to the third floor defines the edge of the double-height volume of the kitchen, and provides a visual anchor separating the floor space into smaller areas. The orientation of this space, relative to the glazing, insures that the middle section of the floor plate receives natural light all day long.

The stair up terminates at a bridge-like element straddling the double-height volume that begins in the kitchen and ends in the dining room. This space was created using the remaining walls of the old elevator shaft. From one side of the bridge, there is a spectacular view towards downtown Toronto. From the other, one overlooks the kitchen and bedrooms beyond. One step up from the "bridge" is the finished third-floor family room. The east/west walls of the third floor are made up of oversized sliding doors that lead out to a deck on either side. The vaulted form of the roof is echoed in the profile of the ceiling. At either end, it drops down to meet a lower, orthogonal bulkhead. These mark the threshold between outside and in. The character of the two exterior spaces was developed to exploit their different orientations. The east deck is more intimate and is finished with uncalibrated slate and built-in planters. The west side is cedar-clad and contains the barbecue and a large seating area.

Janna Levitt

Waverley Hotel

484 Spadina Avenue
Builder, J.J. Powell
Completed 1900

Next to the Scott Mission is the Waverley Hotel, which has played host to the changing population of Spadina Avenue, from periods of Jewish immigration to the hippie days of Milton Acorn to elements of today's Toronto Native community. Originally the site of Robert Milligan's market garden, by the 1870s four small cottages were located on the then pastoral site. The cottages were demolished and replaced by the local YMCA in 1882, which was itself replaced by the Waverley. The building was enlarged in 1925, and by the late 1920s, it housed the Toronto Tourist and Convention Association. After its lounge license was granted in 1955, followed by the creation of the Silver Dollar Room in 1958 (also known as "The Buck"), the Waverley became entrenched as a representative form of single-resident occupancy in the city.

Today, the Waverley Hotel is a 65-room building that rents rooms on a monthly and weekly basis. Daily rates are also available, but the Waverley fits into the category of hotel that speaks of its "residents" rather than its "guests." Once inside the glass doors, there is a large and harshly lit waiting room with a small television and a few smoking souls whose origins and purpose are not evident. Signs abound outlining strict moral codes of conduct as well as precautionary notices. Upstairs, the rooms are dilapidated, yet remarkably clean, and many are claimed by long-term residents.

From 1970 to 1977, one of Toronto's famous poets, Milton Acorn, lived in the Waverley Hotel. He was known for constantly changing rooms because he was suspicious that the RCMP were bugging his room. The Waverley is a form of housing that attracts transients and individuals whose lives are less grounded than most Toronto citizens, and in this respect Acorn fit in with the typical Waverley resident.

EJR

The nondescript building is noted more for its colourful history than for its architectural features. Downstairs, the Silver Dollar Room is preparing for another night of revelry to the sounds of Blues and Rock. The nostalgic murals, redone in 1994, punctuate the bar's tight, no-nonsense space. It's a "straight, no chaser" kind of place. In Ellmore Leonard's novel *Killshot*, reference is made to the Waverley Hotel as a place where men would end up drinking too much, partly because of the Silver Dollar downstairs and partly due to their personal circumstances. He describes the hundreds of light bulbs on the sign outside the bar, and the rooms upstairs with so many cracks in the ceiling that road maps seem to appear for the forlorn staring skyward. The challenge is to find the crack that leads out of the Waverley and out of a difficult life. The Waverley Hotel is an important part of Toronto's architectural history, but of a somewhat less than glorious nature.

Ian Chodikoff

Southeast Spadina

Bounded by Spadina Avenue, College, McCaul, and Queen streets

A curious thing about Southeast Spadina is that it has not gentrified. The neighbourhood is directly adjacent to downtown on the east and the University of Toronto on the north, locales that provide gentrifiers aplenty to other city neighbourhoods. Its streets are treed and pleasant. Many houses need some work, but no more than those in some other inner-city neighbourhoods before gentrification happened. When location and condition are reckoned together, house prices are not out of line. Yet Southeast Spadina obdurately refuses to go the way of Don Vale, Sussex-Ulster, Riverdale, or any of the city's other gentrified areas, all more distant from downtown. It remains a somewhat scruffy neighbourhood that looks pretty much as it did thirty-odd years ago when Albert Franck, Toronto's painter of old houses, plied its streets and lanes making pictures like *Backyard on Baldwin Street*.

A reason commonly cited for the area's stability is the presence of Chinatown. Commercial frontage on Spadina Avenue and Dundas Street is almost uniformly East Asian, and many Chinese homeowners and tenants live on the residential sidestreets. The vivid Chinese presence, it is sometimes said, has been a bulwark against gentrification. Maybe. But the Art Gallery of Ontario and Ontario College of Art and Design are also in Southeast Spadina. Picturesque Baldwin Street Village is there. Just south of the neighbourhood is a district of trendy restaurants, boutiques, nightclubs, and theatres. Nor is gentrification typically a bashful process in the face of reasonable property values a few blocks from a central business district. But it has not happened.

EJR

What nearly did happen was more cataclysmic. The decisive moment in the neighbourhood's recent history was a report published in 1972, *Toward a Part II Plan for Southeast Spadina*, that trenchantly captured tensions in Toronto's politics and planning at the time. Part II plans were meant to frame proposals for applying the City's 1969 Part I plan to local areas. In the case of Southeast Spadina, the Part I plan imagined almost total erasure. Most of the neighbourhood – areas adjacent to the university and the central business district – was designated for high-density institutional and commercial development and the rest for highrise apartments. This reflected the city's prevalent policy in those years of displacing whatever was old and appeared dilapidated in the name of growth and prosperity. Although bourgeois in the 19th century, it became a working-class district after Toronto industrialized and, for decades, was home to Jewish immigrants from eastern Europe (mixed, during and, after World War II, with

EJR

Maritimers coming to Toronto for work). In the 1960s, as the Chinese began replacing the Jews, the City labeled the neighbourhood "blighted."

But Toronto's politics were in ferment at the time – the period when the Spadina Expressway was terminated, when people in Trefann Court and Kensington Market battled to prevent demolition of their homes and businesses for public-housing projects. Throughout inner Toronto, neighbourhoods were fighting the scorched-earth "urban-renewal" agenda rooted in civic boosterism, deformed modernism, and corporate rapacity. Planning was changing, too. The first generation of planners to have been schooled amid debates engendered by Jane Jacobs's caustic critique of modernism, *The Death and Life of Great American Cities*, were now on the staffs of municipal planning departments.

Toward a Part II Plan for Southeast Spadina not only declined to execute the vision of the Part I plan but, in a quiet and measured tone of civic bureaucracy, suggested this vision was dangerous rubbish. Meanwhile, people in Southeast Spadina had been fighting their own battles – against provincial plans for a block-sized electrical transformer at Baldwin and Beverley streets, a Police Department notion for a new division headquarters on quiet Darcy Street, developers' proposals for high-density residential projects that would wipe out two more blocks at Phoebe Street and blanket the east side of McCaul Street. Under the aegis of the new Part II plan, only the last of these was built, in a more street-friendly form than the initial design for a cluster of highrise towers.

Preserved by its Part II plan, Southeast Spadina has marched three decades to the beat of an idiosyncratic drummer but will inevitably gentrify. The seeds are present. While there is not a lot of conspicuous renovation of the old houses in the south part of the neighbourhood, there is already a substantial middle-class presence here, particularly in and around a batch of new houses made to look like Victorian homes built on Phoebe Street in the 1970s; and a luxury-condominium development is underway on Beverley just above Queen. Meanwhile, the northeast quadrant of the neighbourhood, centred around Baldwin Village, will sooner or later yield to inexorable pressure. Only around Dundas, west of Beverley, the main hub of the Chinatown commercial zone, is it likely the process may remain dormant for the foreseeable future.

Wandering the neighbourhood's genial streets today, a passer-by can only wonder how Southeast Spadina might appear if Jacobs had written her book a few years later.

Jon Caulfield

Spadina Avenue residential/ commercial blocks

The mixed residential and commercial blocks that run along Spadina Avenue, from College Street to Dundas Street and more intermittently toward Queen Street date back to the 19th century (1850s -1890s). They are a local example of an old and universal type that crystallized in the Renaissance, when merchants built residences above their establishments, each battling for a slice of street frontage to advertise their wares and services. The Spadina facades are in various states of disrepair today, with instances of Victorian brick detailing concealed under flashings or entire facades plastered over. Despite the

EJR

disrepair and a street smelling at best of fish from the daily offcasts of the Chinese restaurants that inhabit many of the commercial spaces, the Spadina Avenue neighbourhood is a vital and thriving place, much loved by its inhabitants.

The "Spadina Blocks" form this neighbourhood by creating a "wall" on either side of Spadina Avenue, between which a public space is formed. This space creates a concentrated micro-environment, which is able to absorb enriching stimuli from the immediate context: the University of Toronto, the Art Gallery of Ontario, the Ontario College of Art and Design, Little Italy, Kensington Market and the city's downtown are all nearby. This is achieved by a series of devices fundamental to this building type: continuous street frontage, entrances to residences from the front facade, the arcade-like atmosphere resulting from commercial signage, and the ability to concentrate a density of people and amenities. Although the building type contributes to the creation of a quality public space, it also relies heavily on the nature of its context to do so. In this instance, Spadina Avenue plays an important role. It was laid out in 1834 by Dr William Baldwin as a grand, 132-foot-wide avenue (twice the usual street allowance), running from Queen Street to Bloor Street. At the time, a sweep of indiscriminate private-enterprise development was taking place across Toronto, without much public ordering or design and Spadina Avenue was one of the few examples of generous and conscious planning. Spadina Avenue also served as an important streetcar route in 19th century for bringing suburban residents into the downtown area, before commuter train lines were developed. And it remains an important public transportation route.

As Toronto industrialized and prospered in the latter part of the 19th century, it faced the issue of providing housing for its rising population, as the city drew increasing numbers of factory labour and service workers, as well as an influx of immigrants. From the beginning, right up to the present, the Spadina Blocks have housed successive waves of Irish, Jewish, Hungarian, Portuguese, West Indian, and Chinese immigrants. Over time, the residential components of the blocks have undergone intense manipulations, such as the splitting and recombination of apartments, the conversion and reconversion of rooms to serve different functions, the continual upgrading of services, and the swapping of uses from residential to commercial.

The neighbourhood was considered a slum in the 1960s and threatened by renewal, but withstood the pressure. Resilience and adaptability are part of the history of this site, but they are also an important capability of this building type. Today the neighbourhood is again in transition, as it changes from Old Chinatown to a multicultural community of students, refugees from the suburbs, artists, and young professionals.

Marsha Kelmans

Kensington Lofts

50 Baldwin Street and
21 Nassau Street
Current architects, Robert Barnett Architect with Paul Oberst
Conversion completed 1999

The social history of Kensington Lofts is as interesting as the architecture. The site was home to the Provincial Institute of Trades, and then George Brown College until the early 1990s. When the College announced its abandonment of the site, Kensington Market residents, business people, and local politicians, concerned about the future redevelopment of the site, formed a group and developed a project to transform the campus into mixed-income social housing, with a strong emphasis on artists lofts. The community plan dissolved with the failure of the City of Toronto to meet a deadline to get the site at a bargain price in July 1995, and the site was put on the open market.

Hoping to keep some aspects of the original concept alive, the community approached a number of "Kensington friendly" developers and eventually C and A Developments took on the project. C and A, a partnership of Howard Cohen, former head of the Design Exchange, and architect Lloyd Alter, were just completing 20 Niagara Street and liked the community's vision. Without social-housing programs, the project became a market condominium, with retail on Baldwin Street. The developers donated some community space. Over the period of the construction, C and A evolved into Context Developments.

The project is one of the pioneer loft conversions in Toronto. Although aimed at mid market, a high standard of architectural design was achieved. Not only were the development team architecturally trained, the sales team were also architects. The project architects were Robert Barnett Architect in joint venture with Paul Oberst. Robert Barnett Architect had been the consulting architect to George Brown College for many years and so was familiar with the existing buildings. Barnett had also developed an earlier scheme for the Ontario Realty Corporation, which showed how the existing buildings could be adapted for housing. Paul Oberst, a Kensington Market resident, had been involved in developing the early schemes for the site with other community members. Kohn Shnier Architects were commissioned to work on the building elevations. Some of the suite layouts were done by Kohn and Shnier, in collaboration with interior designer Cheryl Krismer and the project architects.

The project has two wings, one on Baldwin Street and one on Nassau. One bay of the Baldwin Street building was demolished to widen the courtyard from 20 to 40 feet. The two wings are linked by a new structure on the east edge of the block.

Architect Alex Spiegel, a partner in Context Developments, wanted the project to be as "green" as possible. The first step was re-using the existing buildings. In addition, a central system supplies heated and chilled water to individual heat pumps in the units. Among the material selections were bamboo, recycled concrete, and glass countertops. The original doors, fitments, and furniture were sold or given away to community agencies. There is a roof garden on the connecting link with Nassau Street, which is visible from the stairwell of the Nassau Street building.

EJR

The high ceilings in the original buildings created dramatic spaces, and the old buildings also allowed for great variety in size and layout of individual suites. The wide school corridors were left as generous hallways. On the second and third floor of the Nassau Street building remain two dramatic coved ceilings from the 1924 school. In many of the halls the services are left exposed, painted dark grey, with lighting suspended below them.

The project sold out very quickly. It took only two years to complete the construction and move in all of the new owners.

Catherine Nasmith

George Brown House: the house

(originally 'Lambton Lodge')
186 Beverley Street
Architects, William Irving
with Edward F. Hutchins
Completed 1876

EJR

This fine Second Empire style house was built for George Brown between 1874 and 1876. The elegance of the residence reflects his prominence as a leading Liberal politician, Father of Confederation, and founder of the successful *Globe* newspaper, now the *Globe and Mail*.

In 1976 the Federal Government declared the house a National Historic Site, but a decade later the house was threatened with demolition. The Ontario Heritage Foundation intervened by acquiring it and converting it to a conference centre with offices on the upper floors. The site is also a popular filming location. Impressive walnut hoods adorn the main hall doors and fourteen fireplaces add character to the spacious rooms.

George Brown's library was recreated by Parks Canada in 1987 and now features 2000 of his books. The restoration of the house focussed on the Brown family period, but later alterations were retained, such as the turn-of-the-century Art Nouveau mahogany paneled dining room. The room had been redecorated by Duncan Coulson, a bank executive who owned the house between 1889 and 1916. Following his death, a school for the blind was built at the back of the house. Forty years later, the CNIB moved out and a school for mentally handicapped children took its place. The school addition was demolished in 1984 and a new school was built on the adjacent lot.

Denis Heroux

George Brown House: the gardens

186 Beverley Street
Layout largely by George and Anne Brown

The land on which George Brown sited his house and gardens constituted about one-third of the parcel he owned along Beverley Street from Baldwin to Cecil. In rough sketches of 1871–75 to his wife Anne, he detailed a cast iron fence enclosing the front garden of grass and shrubs. A private garden with a service yard lay to the west, a conservatory to the north, and beyond that, a kitchen garden and croquet ground enclosed by a high wooden fence. By 1875, "roses and honeysuckles and Virginia creepers" were to be planted around the croquet ground. By 1877 Brown was requesting his wife to consider "shade trees" (chestnuts and rock maples) for the street.

The earliest known illustration – an engraving of Brown's funeral in 1880 – indicates a tall elm and a columnar tree north and south, respectively, of the entrance on Beverley. Ontario Heritage Foundation archaeologists established tree remains that might correspond to the elm, and evidence that there could have been planting beds in the centre of each section of the east lawn. The current front walk, although poured concrete, probably reflects the original form; the diagonal path to the arched gate is a modern addition. Although no plantings found on site are likely to date back to the Brown period, one rose on the west may be significant. All other plantings, including the privet hedge, are recent.

Despite the loss of most of the west and north grounds (and the degradation of the Duncan Coulson verandah, designed in five bays in 1889–90 by architect David B. Dick), initiatives are now under way to create a new garden appropriate to the George Brown period. The restored and rebuilt fence, with its cast iron posts and tracery in a heavy Renaissance Revival style, is the most notable architectural feature to survive from the 1870s.

Mark Laird

Beverley Place (Hydro Block)

15 Beverley Place
Baldwin to Cecil Street,
Beverley to Henry Street
Architects, A. J. Diamond
and Barton Myers
Completed 1974

Beverley Place, also known as the Hydro Block, replaced plans by Ontario Hydro, the provincial hydro-electric corporation, for a 12-storey transformer station in the midst of an established residential neighbourhood.

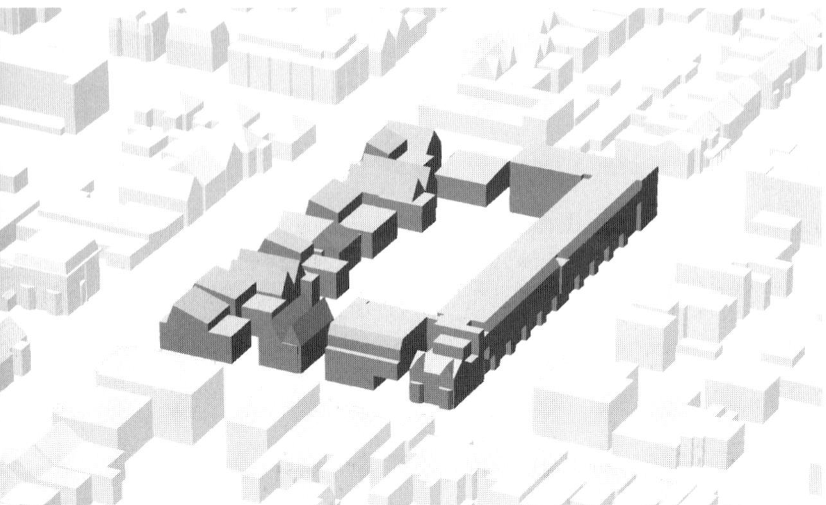

CPD/CT

Following the emphatic objections of the community, CityHome (Toronto's newly formed non-profit housing company) stepped in, acquiring the site and commissioning Diamond and Myers to design a high density residential project that respected the scale and material qualities of the existing neighbourhood. Unlike the architects' earlier Sherbourne Lanes project, not all the existing Victorian homes were saved, as many were structurally unsound; however, the 12 original houses that could be saved were converted into multi-unit residences, and the architects also introduced a single, block-long new building in a Modernist language that by virtue of scale, material treatment, and proportion is successfully integrated into the existing fabric. Beverley Place received a City of Toronto Non-Profit Housing Corporation Neighbourhood Development Award and an Urban Design Magazine Award. In addition to its innovative design, the project is notable for demonstrating how the making of the city is essentially a political process: the architects challenged zoning and development bylaws, and engaged in the vigorous community activism of the day, reflecting a belief in the potential of architecture as a democratizing force, and a view of the city not just as built artifact but as a living, evolving social organism.

Marco Polo

Stinson House

5 Leonard Place
Architect, Jeffery Stinson
Completed 1989

"Think of it as part fortress, part flying machine." "How interesting," they respond, "but what was it BEFORE?" "There's no 'BEFORE," I say, "it's all new."

This three-bedroom family house was built mostly by the architect and his sons on a derelict property in a back lane, in an inner-city neighbourhood. • It was designed for a changeable future, and already it's two apartments and a second redivision of space is about to take place. • The first of the new laneway houses, it uses a tiny lot of waste land and takes advantage of existing city services. • It's designed to maximize sunlight and solar heat and, for economy, it uses conventional materials in unusual ways – inside and out, sidewalk grilles allow sun and air to penetrate. • It's compact, but high

Design Archives: Robert Burley

spaces and a walled garden with a fish pond at living room level make it seem more generous. • The basement has a glass wall opening to a lower garden, so that it becomes livable space. • It makes powerful, but not literal, references to the stable, five-bay, Georgian ideal, to its heroes Berlage and Eames, to the protective strength of armoured helmets, and to the idyll of forest bathing. • It is firmly rooted – it's mass close to the ground, lighter steel and glass above.

Jeff Stinson

Alexandra Park: urban renewal

Bounded by Augusta, Dundas, Cameron, and Queen streets
Architects, Klein and Sears, Jerome Markson, and Webb Zerafa Menkes
Seniors Apartments: Adamson Associates, Architects
Landscape Architects, Sasaki Strong & Associates
Completed 1967–69

It is illuminating for us to look back over 30 years of our architectural practice and review problems, goals and outcomes. In the 1960s this immigrant neighbourhood was designated for redevelopment. Although not a slum, it had degenerated over the uncertainty of its future. Then agreements between the federal, provincial, and municipal governments opened up new possibilities for rejuvenation.

All parties were interviewed and reasonable objectives identified: a small scale; minimal conflict between cars and people; straightforward units; cohesiveness and viability of the neighbourhood; play spaces for children; tough landscaping; retention of trees; a nursery; and smaller commercial spaces.

An 18-acre pedestrian precinct was created to separate people from vehicles, parking was placed at the perimeter of the neighbourhood, and provision was made for emergency vehicles to operate through the site.

Of the 627 housing units, 200 are elderly persons' apartments, 40 are for singles and small families, and the remaining units are in new row houses. These moderate-scale buildings line a winding "north-south main street" spine with branches which encourage cross and through circulation. All family units are at or near grade and have access to gardens. There are a small number of renovated apartments and houses. A serious attempt was made to conserve certain strings of houses, but there were no financial provisions for doing so.

On completion, our team felt that Alexandra Park had been imposed from above and that the residents had insufficient input. We believe that a more gradual and less disruptive process of physical renewal and more rehabilitation of existing buildings would have achieved better results. We wonder whether we should have continued the grid system of roads. Perhaps today we might have added more detail and playfulness to our elevations.

In September 1969, the *Canadian Architect* described Alexandra Park as "a living environment of reasonable density and a high child count, inserted into the worn fabric of the city." Certainly it is identifiable as a project, but anything new inserted into something old is identifiable."

Jerome Markson

Rental housing crisis

More than half of Toronto households live in rental housing. Tenant household incomes (on average, about half of owner incomes) are falling in real dollars, while rents are rising faster than inflation. About one in four tenant households are paying 50% or more of their income on rents, putting them on the brink of homelessness. The income-rent squeeze means that a growing number of households cannot pay the monthly rent. Every week in Toronto, landlords file 500 applications to evict tenants (mostly because the tenants were behind one or two months in rent). On the supply side, new rental construction has collapsed, especially since senior levels of government began cancelling social housing programs, starting in 1993. In 1998, only 114 new rental units were completed, when the city needs at least 2,000 new units annually. The rental vacancy rate is below 1%, which means that there are practically no available units.

Michael Shapcott

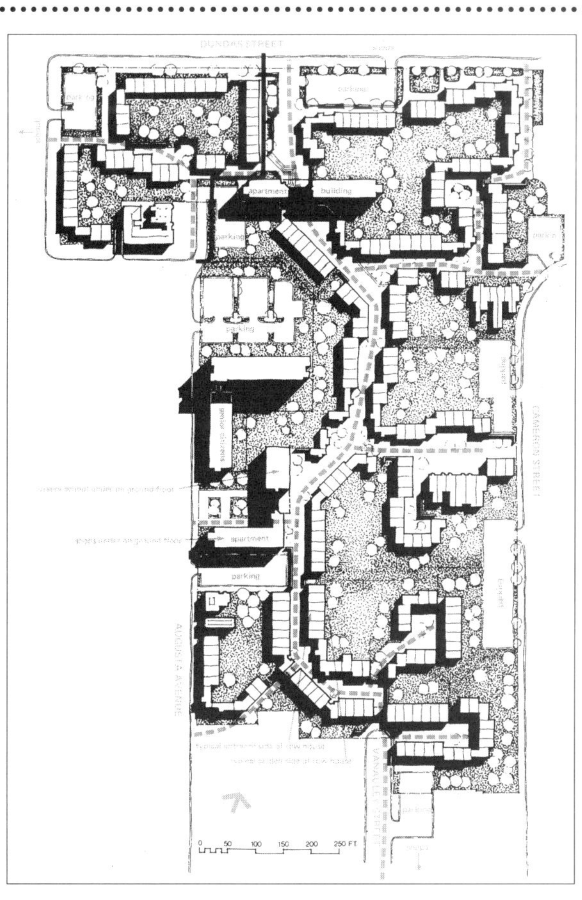

Jerome Markson

15 Larch Street and 76 Grange

Architects, Allen Ensslen Barrett
Completed 1996

Located in Toronto's "Chinatown," the two buildings that constitute 15 Larch Street and 76 Grange attempt to fit in with a dense, low-rise context. Although most of the residential buildings in the immediate surroundings are two-storey, row and semi-detached houses, the form belies the population density as many of the older buildings have been subdivided into smaller apartments. The adjacent main streets are intensely used for commercial and retail activity. Along Spadina Avenue, much of this activity is accommodated in older industrial buildings of up to six storeys, whereas Dundas Street West is characterized by low-rise development similar in scale to the residential stock. The neighbourhood, close to the central business district, is well served by public transit.

SAW

The Larch Street buildings were designed for the Toronto Housing Company (formerly Cityhome), a municipal, non-profit housing provider. Both buildings are three storeys high, fitting with the scale of the neighbourhood. They address the streets by providing many of the units with their own front doors, and the apartments on the east side of Larch are clustered around a semi-public courtyard. The project was constructed on top of a separately owned underground parking structure that was designed to serve the businesses in the area. This complicated the development process, and, despite the establishment of local working groups to discuss the concerns of residents and business owners, the project took almost ten years to complete.

Richard Milgrom

The Grange

Grange Park at John Street
Completed c. 1817

From the Sky Dome, gazing north up John Street, beyond the busy shops of Queen and King Streets, one can see The Grange, serene in its park setting. First home of the Art Gallery of Ontario and Toronto's oldest remaining brick house, this Georgian building was built in 1817 by D'Arcy Boulton Jr for his family.

The Grange was owned by the Boultons for almost one hundred years; and, like any home, it has changed over time. The original house was two storeys, 60 x 40 feet, with a low hipped roof containing a circular window. There were five bays (windows), and the three central ones projected out slightly from under a pediment. Reflecting a sense of Georgian balance, the front door opened onto a central hall with the dining room on the left and the drawing room on the right. At the back of the house and on the second floor were the bedrooms for the family, and there were four large rooms in the attic, probably for the servants. The kitchen, food storage areas, and scullery were in the basement.

Art Gallery of Ontario, Archive N-1098

In the 1840s (the era of the second generation of William and Harriette Boulton), the family enlarged the house by removing the main-floor bedrooms to create a back stair hall and breakfast parlour, and three of the upstairs bedrooms to create an assembly room or ballroom. The main floor was extended to the west for an office. The Grange remained in this form until 1885, when Harriette Boulton's second husband, Goldwin Smith, added a Victorian Library and a new oak staircase.

The Grange underwent two more changes in the 20th century. In her will, Harriette left the house to what was then called the Art Museum of Toronto to be its first permanent home. Opening in 1913, the earliest exhibitions were held in the drawing and assembly rooms. As the Art Museum expanded, The Grange rooms were used as museum offices. In 1973 the next metamorphosis occurred when Art Gallery of Ontario volunteers sponsored the restoration of the house to its earlier glory and The Grange opened as a historic house.

Jennifer Rieger

50 Stephanie Street

Architect, Uno Prii
Developer, Investanti Canada Ltd
Completed 1968

Accounts held at the Building Records office at City Hall show that this project was one, long, drawn out battle. There are 25 microfiche pages on record, 19 of which hold correspondence detailing the struggles of the City and the neighbours against the proposed building and its subsequent construction. This includes Ontario Municipal Board hearings, letters of complaint from the Art Gallery of Ontario, multiple stop-work orders, orders-to-comply, and other pleasantries.

Of interest in the records is a 1965 site plan showing a shorter building, using the same staggered and bent plan, closer to Stephanie Street, along with eight townhouses arranged to the north, next to Grange Park. This plan was not carried out. Instead, when the drawings were submitted for a building permit in 1966, the proposal was for the present 24-storey building only.

The project illustrates Uno Prii's interest in the possibilities of the building type, but unlike his more identifiable buildings that create a singular, swoopy image, this experiment exposes each component of the building to make it as multifaceted as possible. Some of the techniques he uses include:

E/R

- the dog-leg stagger of the plan that plays with the cellular nature of the structural system and its sculptural possibilities;
- the extensions of the concrete shear walls and their punctuation across the facade that fractures the massing of the building;
- the concrete decorative features at grade – the free-standing colonnade and the dimpled concrete canopy (both have the affect of diverting the eye from the facade);
- the cantilevered corner balconies that breakdown the structure by "dematerializing" its corners.

These techniques are used to greater effect on other sites; for example, the shear wall projections of the Elizabeth Street nurses residence or the canopy and fountain at 44 Walmer Road. Here, the inappropriate location for such a tall building makes it harder to wax nostalgic. The building appears apologetic rather than Expressionistic.

Ian Panabaker

Beaver Hall Artists Co-op

29 McCaul Street
Oleson Worland Architect
Completed 1989

Beaver Hall provides a new type of live/work accommodation for Toronto artists, who were otherwise finding it increasingly difficult to find suitable spaces in the downtown area, at affordable rents. A group of artists formed a Housing Cooperative and bought the site, at a central location just north of Queen Street West, south of the Art Gallery of Ontario and the Ontario College of Art and Design.

All units have high ceilings and large windows. The bathrooms and kitchens were located compactly near the rear of each unit, permitting an open plan at the front. A flexible layout was assured as the building was designed so that there was no need for loadbearing partitions carrying plumbing pipes and electric conduits. Tough interior surfaces – polished concrete floors and exposed concrete block walls met the artists' need for durable finishes.

The typical floor plan has four units around a compact central core, 24 total units. Each unit has an outside corner with wrap-around window and a balcony. There is an art bookstore and an exhibition gallery/workshop on the ground floor. Parking is off site.

Named after a legendary artist's co-op in Montreal, Beaver Hall showed that it was possible to build a medium-density, multi-unit residence on what had previously been considered an "unbuildable" urban lot, and it serves as an economical prototype for artists, demonstrating the concept of "live/work" zoning. Forward-looking – it anticipated the trend in loft-apartments in the city.

David Oleson

• Oleson Worland

The Kings: a bold move

King/Parliament and King/Spadina

In the mid 1970s Toronto took the radical step of resisting the expansion of its office core and encouraging residential buildings and mixed office/housing projects to be built throughout the downtown. That heterogeneity is one of Toronto's great successes, and has been an important part of making it a livable, safe, and interesting city. The strategy known as "*the Kings*" may prove to be equally significant for Toronto's future.

The Kings are two, very large former industrial districts covering approximately 500 acres on both sides of the city's downtown. From the turn of the century to the 1950s, they contained much of the city's industrial base, but industry has dispersed globally and to the expressway-served open spaces at the city's edge, leaving the older inner-city areas under-used and decaying. Yet, these areas are within the orbit of the heart of the city, a prime location, and they contain a fine selection of sturdy, unpretentious, old warehouses, some of which have great historic character.

Two areas, known as King-Parliament to the east and King-Spadina to the west, had been deliberately preserved by City policy for industrial purposes, to protect blue-collar jobs. The policy was perhaps a little unrealistic, but it had the effect of saving the two areas from the development boom which ended in 1989. Had office development expansion been allowed in the 1970s and 1980s, much of the existing building fabric in these areas would likely have been removed. The policy also had the desirable effect of keeping the city's financial district tightly concentrated.

During the recession of the 1900s, these inner-city industrial areas began to feel even more down at heel. One saving grace was that in the early 1980s the City had allowed night clubs, large bars, and other entertainment places to locate in these areas, because they were considered too noisy for other neighbourhoods. People had also begun to use the warehouses as residential lofts although it was illegal to do so; there was a need for cheap housing and buildings were empty. In this way, the Kings organically took on a bit of an anarchic and interesting character: the New York idiom, but not yet part of Toronto's urban vocabulary.

In the mid 1990s, the City's planning staff, under the leadership of Paul Bedford, promoted a bold new idea for the Kings. There was little new demand for offices, and the industries continued to move out for a variety of reasons. To get development going again as Toronto slowly emerged from the recession, the two areas were redesignated as "*reinvestment areas.*" Building and land owners are now allowed to use their existing buildings for any use, or combination of uses, that they choose (with the exception of specific toxic industries). Owners can change the use to which a building may be put, at any point in time, in response to the market. The rules for new buildings relate only to the height of the building and the building envelope, to make sure street relationships and daylight penetration are appropriate. The scale of new buildings is sympathetic to the existing ones, and the few new rules are tilted somewhat towards encouraging owners to save the existing buildings.

The Kings represent the first large-scale attempt in North America to let a new type of neighbourhood emerge largely through the play of market forces, tempered by some relatively simple ideas about city form and scale. There is also an open-ended commitment to meet new social needs when and if they arise and possibly in unconventional ways.

This radical new policy has been amazingly successful, particularly in creating new housing. More than 50 housing projects have been started, an almost equal number of conversions and new buildings. The two areas are coming alive beyond anyone's expectations. It is almost certain that the next wave will be the conversion of warehouses to high-tech offices, with new high-tech companies seeking both the image of the converted warehouse as well as the large group of younger workers who like to live in the heart of the city. Cafes, restaurants, other forms of entertainment, and a wide range of everything, are starting to find their way into these two neighbourhoods. No one knows quite what is going on because the process is no longer mediated by the formal public approval process.

The first round of new buildings ranges widely in architectural quality: from really interesting architecture by Peter Clewes, through the larger body of nondescript buildings, to the sad, suburban apartment form imported into downtown. The future looks exciting as two new neighbourhoods take shape in a manner better than anything we might have imagined, and different. Indeed, the two neighbourhoods are likely to splinter into five or six smaller neighbourhoods, each with a fiercely distinct and different character. Stay tuned as Toronto continues to reinvent itself; it's going to be interesting.

Frank Lewinberg

Camden Lofts

29 Camden Street
Architects, Oleson Worland
and CORE Architects;
Interior Design, Celluni Simone Inc.
Completed 1999

A new alternative in residential accommodation, Camden Lofts fills a former parking lot in Toronto's "Fashion District." The project provides 59 live/work units in an eight-storey, mid-rise building, realizing the promise of the City of Toronto's Reinvestment Area initiative in the King/Spadina area.

Although the building is all new construction, it has many of the attributes of loft-type buildings originally constructed in the area for the garment industry, including high ceilings (10 feet), open plans (partitions at seven-foot height, open above, large sliding doors between rooms), and large, operable windows. Exposed concrete floors, walls, and ceilings provide flexibility and durability within the units, which are accented by industrial details (lighting, trim, fixtures, etc.).

● Oleson Worland

The street facade is clad in brick complementing adjacent structures along the street. The top is articulated by set back floors, which provide exterior terraces for two-storey penthouse units. The circulation corridor is asymmetrically positioned on typical floors, with more (deeper) units facing the street, and fewer (shallower) units facing the mid-block courtyard.

Camden Lofts is an important prototype for mid-scale infill, that strengthens and reinforces the urban fabric and demonstrates the effect of enlightened planning and zoning policies in the King/Spadina area. The project is an alternative model for apartments for the live/work market.

David Oleson

The Phoebe

1 Phoebe Street, 5–29 Soho Street,
12–34 Beverley Street
Architects: Burka Architects
with Wayne Swadron
Expected completion, 2002

The redevelopment of the property bounded by Soho, Phoebe, and Beverley streets is an expression of downtown urbanity in what was once a part of a fine-grained 19th century village. In the 1790s, this site was part of a Park Lot Crown grant to Family Compact member Peter Russell. Over the next century and a half, the grant lands were subdivided and a neighbourhood evolved. North of Queen Street, shops and a jumble of housing and industry were built cheek by jowl. South of Queen Street more industry and warehouses sprouted.

• Bronwyn Krog

Over the 19th and 20th centuries, the Soho, Phoebe and Beverley block acquired a foundry, a chemical works, a livery, a row of semi-detached homes, a warehouse, a church, a paper company, and offices. The Weston "Model Bakery" (c.1897) was on the north side of Phoebe.

Today the area remains varied with mainly retail, offices, and a 1990s infusion of high-rise housing, although the pattern of redevelopment is at a larger scale. Liberalized zoning has contributed to a renaissance in reinvestment in the area.

"The Phoebe" will be an entirely residential project of 230 suites on two acres. In order to fit into the neighbourhood, it is designed as three separate buildings, each of a different size and aesthetic expression, with parking out of sight, underground. The architectural style relies on the "Victorian Industrial" feel of the area, with expansive glazing and brick detailing. The buildings are organized around a tranquil courtyard, and the extra-wide Soho Boulevard, a legacy of the Park Lot subdivision, will be landscaped.

Bronwyn Krog

District Lofts

388 Richmond Street West
Architects, Wallman Clewes Bergman;
Project Architect, Peter Clewes
Expected completion, 2000

It might seem a bit premature to include in a guide to Toronto's housing, a building that will not be completed before its publication, but doing so might not be unreasonable at a time when new building types and approaches are emerging with startling rapidity. District Lofts is one of the most adventurous of the many new residential buildings that have been made possible by the declaration in 1995 of a "Reinvestment Area" south of Queen Street. The relaxed zoning made it feasible to build on an otherwise unpromising centre-block site in the dense warehouse district just east of Spadina Avenue.

EJR

The architects have responded to the difficult site with a parking-podium solution that specifies three levels of parking for residents below grade and three more public-access parking levels above grade. Although the parking garage may not improve the street, which is one-way and carries fast and heavy traffic, it will elevate most of the units above the 401 Richmond building that stands directly to the south, thereby improving views and exposure. The building is organized as two narrow slabs parallel to the street, forming an interior courtyard. This creates a condition where half of the units will not have an orientation to a major street, and it is hoped that the dual exposure of the two-storey through-units that begin at the seventh floor will compensate for the increased enclosure. The circulation in the building is more complex than the Twenty Niagara condominium by the same architects; here 148 units are served by the same number of elevators that serve 21 units at Twenty Niagara. Exterior bridges will span the west side of the courtyard linking the required exits. The courtyard itself, begins on the sixth floor, above the parking and communal facilities, but it will be eight stories deep, something quite untried in Canadian residential building.

Kenneth Hayes

Clarence Square and Clarence Terrace

5–16 Clarence Square
Completed 1879–80

Clarence Square was designed in the 1830s when the Military Reserve was opened for residential development. The Reserve was a large tract of vacant land that surrounded Fort York and went as far east as Peter Street and as far north as Queen. Wellington Street, the centrepiece in the development plan, was laid out as a broad residential boulevard between two classic squares – Clarence

EJR

Square to the east, and Victoria Square to the west. In the grand scheme, the Governor General's mansion was to be erected on Clarence Square and surrounded by large villas of the well-to-do, but the entire development faltered for both political and economic reasons and very little was built. The introduction of the railways in the 1850s made the area even less desirable for residential use and it wasn't until the 1870s that building construction picked up. Clarence Terrace, with its mansard roof, is a rare example of an 1870s rowhouse. Once a very common building type, most have now disappeared. The stucco, which was added during the 1960s when the buildings were rehabilitated, gives the terrace an uncharacteristic appearance.

Michael McClelland

Twenty Niagara Condominium

20 Niagara Street
Architects, Wallman Clewes Bergman;
Project Architect, Peter Clewes
Completed 1998

Twenty Niagara offers such thoroughly integrated advantages for new inner-city residential development that it could well revolutionize Toronto's urban planning, but it might still prove unable to stem the flood of new townhouses that imitate what the nineteenth century left undone. At six storeys (eight if one counts the partly sunken parking garage and the penthouse mezzanine), the building provides the density the city needs, while at the same time its organization preserves the domestic values Torontonians hold so dearly. Each of the 21 units has a dual aspect that not only allows through ventilation and light penetration, but also preserves the spatial variety and clarity of orientation characteristic of a townhouse. The park face has two elevators at its quarter points, each serving two units per floor. The tiny entry lobbies above grade are still public spaces, and thus need two means of egress. The architects have devised an ingenious solution: an electromagnetic control (connected to the building's fire alarm) to unlock the unit doors in an emergency, thereby giving access through either interior to a narrow balcony on the building's west face, which leads to exit stairs at the north and south corners, much like a traditional fire-escape. In effect, the architects have up-dated the traditional city walk-up to meet new code requirements and life-style expectations. The architectural benefits are profound; the efficiency of the corridor-less plan even results in lower monthly costs. The building's urbanity is demonstrated by its response to its park-side site. The top of the parking garage forms an elevated terrace along the edge of the park without impinging on the public; instead, one feels the presence of a fresh and distinctly urban attitude.

Kenneth Hayes

E/R

The Railway Lands

South of Front Street,
between Bathurst and Yonge

The Railway Lands are a 200-acre wedge of land that has separated downtown Toronto, south of Front Street between Yonge and Bathurst streets from the waterfront for nearly 150 years. After the rail companies determined in the 1960s that the lands south of the main corridor were no longer needed for rail purposes, their real estate divisions developed a number of plans for the redevelopment of the lands. The ownership pattern was complex, but generally, Canadian Pacific owned or controlled the lands east of John Street, and Canadian National owned the lands west of John Street.

The first plan, developed in the early 1970s and called Metro Centre, would have replaced Union Station and created a modernist multi-levelled, high-density office and residential development. Of this vision, only the CN Tower was built. The second plan, developed in collaboration with the City of Toronto in the 1980s, proposed a pattern of streets, blocks, and parks reconnecting the city to its waterfront, on which high-density office (generally in the east) and residential development (generally in the west) would occur. Of this vision, the Skydome, Railway Lands Park (and convention centre expansion), Bremner Boulevard, and parts of the street plan east of Spadina Avenue, were built in the late 1980s and early 1990s. A third plan, essentially an enhanced version of the second plan, was completed in the early 1990s, and it imposed greater certainty on the form of development.

In the mid-1990s, Canada Lands (successor to CN) sold its lands to Concord Adex. Concord Adex sought permission to develop a fourth plan for the area west of the Skydome, based on the residential building typologies (low-rise buildings and point-towers) that it had successfully used on the Expo lands in Vancouver. After negotiations between the City and the developer, and a series of public meetings, a new plan was developed and approved in late 1998, which would create a pattern of streets, blocks, and parks, framed by street-related buildings, generally four to six storeys in height, punctuated in appropriate locations by residential point-towers. These changes were not without controversy, especially over matters of building height.

As of winter 2000, Concord Adex had commenced work on their first two buildings (designed by James Cheng of Vancouver and Page and Steele of Toronto) and had finished design work on the next two (designed by Peter Clewes of Architects Alliance) on the south side of Front Street between Spadina Avenue and Blue Jays Way. All four buildings are unapologetically modern and may represent the beginnings of a significant shift in style, as well as type, for Toronto's residential architecture. The next new construction is expected first in the area between the Skydome and Spadina Avenue, followed by the development of the area between Spadina Avenue and Bathurst Street. In the meantime, Concord Adex has built a temporary golf course on the lands between Spadina and Bathurst which will open in 2000.

Robert Glover

Housing on the central waterfront

South of the Gardiner Expressway, from Spadina Avenue to Jarvis Street

With the advent of the age of rail, Toronto temporarily lost its battle for a refined waterfront to the forces of heavy industry. Gone were the city's early 19th century plans for a green promenade, "the Esplanade" on the water's edge, replaced, by rail spurs, warehouses, roundhouses, marine terminals, industrial slips, and in the 1950s, an elevated highway. In the 1970s the tides turned: 1972 saw the creation of a crown corporation that was handed some harbourfront lands by the federal government and charged with the mandate to develop an urban, waterfront park. In this public development, housing was a minor feature, appearing only in the conversion of the Queens Quay terminal to a mixed use of shops, offices, and four stories of luxury condominiums – a project by Zeidler Roberts Associates.

The first significant injection of housing into the central waterfront came not from the public, but from the private sector. The best model of urban waterfront development creates a significant and continuous public space along the water's edge separated from private development by a public street. Ruefully, developers and their architects erected a wall of buildings around the foot of Bay Street, beginning with the Harbour Square condominiums and Harbour Castle Hotel in the early 1970s and culminating with One York Quay by Clarke, Dowling and Downey Architects on the York Street Slip in the 1980s. The Harbour Square development between the York and Yonge Street slips was constructed directly on the water, on the south side of Queens Quay. Buried behind this wall of buildings and invisible from Queens Quay is the Toronto Island Ferry Terminal, from where over a million people begin their visits to the Toronto Islands every year. In addition to the Islands' visitors, a resilient community of about 450 people who live on the Islands use the Ferry Terminal regularly.

Subsequent housing achievements on the Harbourfront between the Spadina and Jarvis Street slips are, with few exceptions, dubious. The three-tower apartment complex constructed at 250–70 Queens Quay (north side) is affectionately known as the "three ugly sisters." Not to be outdone, due west from the sisters, at 350–390 Queens Quay, are a pair of residential towers that

SAW

SAW

have earned themselves the moniker, the "two ugly brothers." Between these two complexes sits a vacant site, 316 Queens Quay, which is owned by the City of Toronto and is due to be developed into a public park. One past housing project with a built form that is more sensitive to its waterfront location is Harbour Terrace at 401 Queens Quay (designed by Daniel Li Architect). Though it sits, like the Harbour Square buildings, on the "wrong" side of Queens Quay, the scale and treatment of the building make for a less obstructive presence on the water's edge.

SAW

There has been some success recently in the generation of new public space. The publicly owned Harbourfront Centre controls 10 acres of public space and facilities. Public transit secured a firm foothold in the area with the introduction of the Queens Quay LRT in about 1990. This important link to Union Station and the subway now includes a connection to the Spadina LRT, and soon to Bathurst Street, Ontario Place and the Exhibition grounds. Yet, the Achilles heel of the area remains the lack of a generous, continuous pedestrian connection along either the water's edge or along Queens Quay, whose streetscape quality is inconsistent. Establishing and improving connections in the public realm are essential if the visitor's experience to the central waterfront is to have some coherence.

SAW

New residential projects include, at time of writing, proposals for 200 and 226–230 Queens Quay by Kirkor Architects. Like their predecessors, the designers will have the task of creating buildings that are both open and oriented to the streets and waterfront, and city views to the north, but suitably protected from the noise and pollution of the Gardiner Expressway. In addition, a new condominium designed by Graziani and Corazza Architects is due to replace a modernist building at the northeast corner of Spadina Avenue and Queens Quay. Giving these projects names like "The Riviera," and "Aqua," and offering an amalgam of worldwide waterfront architectural styles the developers hope to capitalize on the attraction of waterfront living, albeit on a waterfront that is neither the temperate Mediterranean or Floridian scene such names or architectures attempt to evoke. The new developments east of Spadina Quay bear testament to the continued segregation in Toronto of private and affordable housing. Affordable housing on the waterfront has long been relegated to the Bathurst Quay at the foot of Bathurst Street.

Lewis Poplak

Harbourfront West

Spadina Avenue to Bathurst, south of Lakeshore Boulevard West

The form and density of Harbourfront west was negotiated after the federal development freeze of the late 1980s. The new arrangement prohibited development south of Queens Quay West and to the north, set height limits at 12 stories. Much of this land between Spadina Avenue and Bathurst Street remained undeveloped until 1995, when the housing market in Toronto began to rise from the ashes.

Kings Landing

480 Queens Quay West
Architects, Arthur Erickson with John Cowle
Completed 1987

SAW

One development that escaped the federal freeze, this luxury condominium development is set apart from the point towers to the east by its south-facing terraces and solariums, and an early 1980s fascination with green glass. Penthouse units in this building sell at a premium. The extensive use of the solarium in Toronto was taken to extremes at Harbourfront, because solariums were not considered part of a unit's gross floor area if unheated; yet they are easily converted into livable space. The ground floor of this building was renovated in 1994 by Kuwabara Payne McKenna Blumberg Architects to accommodate the National Ballet of Canada practice and rehearsal studios. This large area was originally proposed as a federal technology/entertainment centre and storefront retail, but neither was realized.

Yo Yo Ma Music Garden

Designed by Julie Moir Messervy and Yo Yo Ma
Completed 1999

Across the street, the Yo Yo Ma Music Garden designed by Julie Messervy, a Boston garden designer, was opened in 1999. It is inspired by the interpretation of the Bach Concerto by its namesake, a world-renowned cellist. Be sure to venture in and experience all five musical movements of the garden, each featuring public art commissions, including an ornate wrought iron bandshell and a wind-propelled maypole. All the furnishings for this garden were custom designed. In the summer, on Sunday afternoons, you can enjoy musical performances by the Toronto Symphony.

500 Queens Quay West

Architects, Kuwabara Payne McKenna Blumberg
Completed 1999
This private luxury condominium development is distinguished by its "L" shape, unique set-back conditions, and use of brick, a material that is uncommon in the new Harbourfront buildings. Unlike many other

SAW

Harbourfront buildings to the east, the north facade is a delight from the expressway. The extensively glazed and much coveted penthouse units include two-storey lofts, which afford exclusive views to Toronto Bay and the city.

Queen's Harbour

Completed 1998
Originally slated as the site for the Harbourfront Fire Hall, these condominiums entered the private market in 1997 at a price considered "affordable." The white, quasi-nautical themed structure sits at the mouth of the buried Garrison Creek. By Monarch Park/Urbancorp, this development was one of the first in the wave of mid-1990s construction.

Leslie Woo

Bathurst Quay

Bathurst Street to Stadium Road, south of Lakeshore Boulevard West

The Bathurst Quay Neighbourhood sits upon the site of Toronto's demolished Maple Leaf Baseball Stadium, which had its heyday in the 1950s and 1960s. Unlike most of the development to the east, this neighbourhood is composed of affordable housing. Arranged around the five-acre Little Norway Park (a tribute to the Norwegian Air Force that barracked here during WWII) are two City-owned-and-operated properties (rent geared to income), four housing co-operatives, and one affordable market-housing building. This neighbourhood suffered a dire lack of community amenities for almost 12 years, until 1997, when the community centre and public school were completed on the east side of the park. The neighbourhood is overshadowed by the Canada Malting Silos, which are to be renovated into a national music centre, with studios and performance spaces.

The residential buildings on Bathurst Quay are more significant for the diverse contributions they make to Toronto's social housing stock than their design innovation.

Harbourside Co-op

Fleiss Gates McGowan Architects

The co-op combines 34 townhouse units, each with full basements, and 21 apartments on the third level, all arranged around a central courtyard.

696 Queens Quay West

Architect, Charles Simon Architect

SAW

This is a complex of "garden" apartments, which are double stacked and coated in grey stucco, and a high-rise apartment building.

Winward Co-op

34 Little Norway
Architect, Roger du Toit

SAW

This building has won awards for the high level of barrier free access it affords in all units. Its interior courtyard has a raised garden for the enjoyment of wheelchair tenants.

Harbour Channel Co-op

Corner Stadium and Lakeshore Boulevard
Completed 1990
Architect, Paul Reuber

The most interesting aspect of this building is the story of its genesis. To gain more air rights in the central core, the Bank of Nova Scotia (for its Scotia Tower) made a contribution to the Toronto Co-operative Housing Federation, in trust. The City then leased the land to the Co-op for the construction of the housing. The building also has single-mother units specifically for a non-profit organization called Jessie's.

Leslie Woo

Arcadia Co-op

680 Queens Quay West
Architects, A.J. Diamond, Donald Schmitt & Co.;
Project Architect, Andre Lessard
Completed 1985

Arcadia Co-op is an example of the specific needs community housing realized under the provincial co-op legislation that was swiftly eliminated by the Conservative provincial government after its election in 1995. The design was the product of an intensive consultation process with artists, organized through the Toronto Federation of Housing.

The eight-storey building contains 94 units, arranged in two wings joined by a cylindrical element on the east corner. The ceilings are three metres, higher than normal, but not tall enough to make true lofts, except on the north face of the building, which has two-storey units with tall windows and a full mezzanine. The apartments have a few unfinished concrete surfaces, some occupant removable/replaceable walls, and extra storage. Some of these features might not be so much specific needs as widespread desires; in this sense, Arcadia might be described as a precursor to the domestication of the loft that took place in the 1990s. Although rent remains geared to income, the general sense of the place is that it houses upwardly mobile workers who work mainly in Toronto's film industry.

SAW

The unsatisfactory grade relationship on the south face has a stilted quality. The lakeside site is attractive but isolated, and almost seems to demand car ownership. It also means that the building itself has to constitute the community, which it does with numerous collective facilities, including a members' gallery in place of the usual common room, but whether outsiders often visit the gallery is not clear. New park facilities and the recent Bathurst community centre by Patkau Architects are an improvement for the whole district, where public housing was built to atone for the development sins of the 1980s.

Kenneth Hayes

(distant)
166
158
NORTH TORONTO
Bathurst-Eglinton
Cedarvale
Spadina
Avenue
Chaplin Estates
South Eglinton
Leaside
156
Millwood
154
Forest Hill
Forest Hill Village
Humewood
Deer Park
Moore Park
Bennington Heights
Governor's Bridge
DVP
Casa Loma
South Hill
Wychwood Park
Summerhill
Rathnelly
North Midtown
Rosedale
Pape Village
Seaton Village
Annex
Bloor/Yorkville
Bloor-Bathurst-Madison
Yorkville
Parliament
Playter Estates
Greektown on the Danforth
Danforth by the Valley
Sussex Ulster
University of Toronto
Palmerston
Little Italy
Harbord Street
St. James Town
Riverdale
Cabbagetown
Old Cabbagetown
Little Italy
Kensington
Chinatown
Downtown
152
Gerrard
Elm Street
Regent Park
Leslieville
Trinity Bellwoods
Alexandra Park
Grange Park
Yonge/Queen-Dundas
Jarvis
Moss Park
Trefann
Broadview
Pape
Queen/Broadview
Corktown
Niagara
King/Spadina
St. Lawrence
Town Of York
King/Parliament
St. Lawrence
Railway Lands
Bathurst Quay
Harbourfront
150
Ward's Island
Algonquin Island
148

Toronto Island Community

Each of the seventeen or so islands known collectively as "The Toronto Island" has its own, distinctive history. Even the two remaining residential areas, Ward's and Algonquin Islands, have very different pasts.

Algonquin Island was developed as a result of the creation of the Toronto Island Airport in the late 1930s, when about thirty of the summer cottages that had stretched along West Island Drive were floated over to what had been called Sunfish Island and placed around the perimeter of the newly named Algonquin Island. Most of the remaining houses were built after the Second World War as year-round residences. The post-War bungalows and the earlier West Island cottages with their distinctive battered skirts and miniature mansard windows are still easy to distinguish on Algonquin.

Sally Gibson

Ward's Island started residential life as a camping ground. Known as "tent city," life on Ward's was necessarily informal and fairly primitive. For example, there was no electricity, no refrigeration, and no clean "City" water until 1906 when communal, cold water taps finally sprouted. Early campers had to light lamps, dig holes, and erect little windmills to solve their utility problems.

Each Victoria Day, mainlanders ferried their camping gear across the Bay to "set up," i.e., erect their tents on wooden platforms arranged along sandy streets. Tenters were naturally careful to keep cooking fires away from the cloth tents, but over time, canvas runways were erected to protect “the cook” and others when dashing between living and kitchen areas.

Eventually, the rustic life under canvas began to pall. In the late 1920s, Ward’s Islanders started lobbying the City for permission first to erect wooden roofs, then entire buildings, on their campsites. When the City refused, some Islanders found ways around the restrictions. “You weren’t allowed to close the tent in,” long-time tenter Daddy Frank Staneland revealed. “This was canvas. But a lot of us fooled them. We put the canvas on the outside and then had wood inside at each end.”

By 1931, the Ward’s Island Association finally persuaded City Council to allow Ward’s Islanders to build permanent cottages on their campsites. Each one had to contain no more than 840 square feet, had to have electricity, and had to be “maintained in a state satisfactory” to civic officials. Even before the final regulations were passed, Ward’s Islanders pulled out their hammers and got to work on their “permanents,” as they fondly referred to them. A thriving cottage industry soon sprung up: Sheppard and Gill Lumber Company, architect Bob Maginnis, and even the Robert Simpson Company got into the act. Two years later some residents were already looking back nostalgically to the good old tenting days.

Today, Ward’s Island still reflects its tenting origins. The pedestrian streets follow the old pedestrian pattern; the tight spacing of the year-round houses follows the distribution of the old campsites; and many of the surviving houses follow the basic lay-out of the old canvas tent-cum-kitchen-shed living spaces. The communal spirit of the early tenters also lives on.

Based on Sally Gibson’s More Than an Island: A History of the Toronto Island. *Toronto: Irwin Publishing Inc., 1984.*

Sally Gibson

Ward's Island

The sandy beaches, soothing breezes, and lush lagoons of the residential community on Toronto's Ward's Island are worlds apart from the chaos and insanity of the city's downtown core. Surprisingly, access to this idyllic paradise is only a ten-minute ferry ride across the harbour.

The Ward's Island landscape of picturesque cottages on narrow lanes lined with trees, flowers, and vines appears to be an eccentric garden rather than a residential neighbourhood of 700 people. Like a bag of Licorice Allsorts, the cottages vary widely in colour, texture, and design. But the commonality of their modest profile, appealing dormers and diminutive 40'x45' lots, tie them into an integrated whole, just like the residents of the island.

• John Ota

"We love living here," says Ruth Howard, who, along with her husband Steve Cooper and their children, have recently built a new house in the island community. "We're living in the city but we're also living in the country. We're fortunate to have the best of both worlds." The family was able to build on the island when, in 1996, the Toronto Island Trust offered 12 building lots for sale on Ward's Island. The lots were awarded by lottery from a waiting list of over 300 hopeful applicants. Howard and Cooper were among the ecstatic winners.

Their two-storey, 1,300 square foot house was designed by the husband-and-wife architect team of Marjut and Klaus Dunker, who creatively used the Toronto Island Zoning Study to help fit the new building into the island's pastoral setting. According to the architects, the study provides guidelines that maintain the existing scale of houses on the island by limiting the size of floor plans and encouraging low lying roofs and asymmetrical facades. "The surrounding community has an evolving self-built environment and we wanted to express that characteristic in the design," says Klaus Dunker. "We hope that the building will blend into the landscape and eventually will look like it has been there all the time."

Indeed, the new house with its clean lines, vertical wood siding, and multi-window facades, is a lively re-interpretation of the surrounding structures. The different planes of the front facade slide back and forth to minimize the visual impact of the home. To further meld their dwelling into the island atmosphere, Ruth Howard matched the muted tones of blue, green and red of the existing buildings to their own house exterior.

" The architects did a great job with the look of the house," says Howard. "It's a cottage with the traditional island look, but in a modern style. It's exactly what we wanted." The main floor is an open concept plan that includes the kitchen, living, dining, and play area. "The kitchen is the centre of the house and the place where we spend most of our time," says Steve Cooper. "In the kitchen we play Scrabble, talk, and just enjoy being together."

On the second floor are the sleeping areas, bathroom, and storage areas. Interior finishes are simple, including natural wood and bathroom tiles that were custom-painted by the children. The 75 windows of the house flood the interior with natural light and provide views outside that give the feeling of living in a tree house.

Ecological concerns were also incorporated into the design of the house. Steve Cooper proudly points to the large south facing windows and concrete floor that absorbs heat from the sun and passively warms the dwelling. Another conservation strategy was to build parts of the structure with recycled wood beams that the family obtained from a former barn. "The whole idea was not to cut down more big trees," says Cooper.

Life on the island is characterized by a laid back casualness combined with deep-rooted community spirit. In the early morning, residents quietly tend their gardens and easily float from house to house for visits and conversation. "We've been lucky to live a different lifestyle over here," says Ruth Howard. "The neighbours are very friendly and we walk to the beach and enjoy all the gardens."

In a city that is struggling for direction after amalgamation, the hallmark of Toronto remains the diversity of its neighbourhoods, where people of all spots and stripes can find their place. With its funky cottages, bucolic aura, and community panache, the island is a unique neighbourhood in the city and the closest thing to being on vacation all year around.

John Ota

Coxwell Stables – CityHome Housing

Coxwell at Gerrard Street East
Completed c.1985
Oleson Worland Architects

This project involves the renovation, and adaptive reuse, of two historic stables in the east end of Toronto. The municipal stable complex dates from the time when the Public Works Department was "horse powered." When the structures were declared surplus, it was proposed that they be renovated into housing for the City Housing Department.

Nine two-storey, two-bedroom townhouse units were constructed in the main building – the width of one four- horse stall yielded one residential unit. A one bedroom and a bachelor unit were constructed in the separate office building with common facilities – laundry and storage. The site development takes advantage of its location next to a public park. Since the buildings were listed by the Toronto Historical Board, a particular sensitivity was required to retain the exterior character of the buildings.

● *Oleson Worland*

The cost of the project was below the Maximum Unit Price set by the provincial funding program, proving the economic viability of renovation. Coxwell Stables has received awards from Ontario Renews and the Toronto Historical Board.

David Oleson

Candy Factory Lofts

993 Queen Street West
Quadrangle Architects Limited
Completed 1998

This conversion of a historic factory into residential lofts is the first large-scale loft development to be launched in the city. It features classic mill construction, huge windows, and a variety

MWF

of unit types and sizes. The project involved creating a new sixth floor as a penthouse level and terraces for the fifth floor units, as well as a complex construction process to make two levels of underground parking below the building. The Candy Factory Lofts have been the impetus for on-going redevelopment in the neighbourhood.

Leslie M. Klein

Garden Court Apartments

1477 Bayview Avenue
Architects, Forsey Page and Steele
Landscape Architects, Dunington-Grubb and Stensson
Completed 1939–41

The Garden Court Apartment housing development is one of the best examples in Toronto of skillful integration of architecture and landscape architecture. The project, built between 1939 and 1941, was the result of the close collaboration of architects Forsey Page and Steele with Dunington-Grubb and Stensson, the city's best-known landscape architectural firm of the period. The property covers a 5.5-acre site located in Leaside, then a growing suburban residential area some 3.5 miles northeast of Toronto's central business district.

William N. Greer

The design of the block plan is axial, with the apartment buildings grouped around a large, central, landscaped courtyard, extending from Bayview Avenue to Berney Crescent and secluded from passing traffic. Along the central axis, the site's slight downward slope from west to east allows the courtyard to be connected by short flights of steps. In the central space there is a grove of lawn trees (now ash and linden; originally also white elm). Along the main axis, each area has its own palette of woody plants. An espaliered fruit tree (one of an original pair) is located near the west end, as is a row of eight Camperdown weeping elms on each side of the walkways. Similarly, towards the east end, a row of weeping mulberries flanks the walkways. Whether approaching on foot through the entrance arcade from Bayview, up the steps from Berney, or through openings in cedar hedges lining service drives at the north and south, the pedestrian is guided by flagstone and asphalt walks lined with trees and hedges.

The outer facades of the buildings face less formal, smaller courtyards, formed in the four corner areas for recreational use. The original plan shows that these spaces were planned for either tennis courts or children's playgrounds. The north tennis court was the only one actually built but it was removed about ten years ago.

Vehicle access to the site for deliveries and parking is provided only from the east and west streets to lanes on the periphery. A continuous row of garages defines the north and south boundaries of the site.

Garden Court's "beautiful park-like surroundings" were described in 1939, when many of the units first went on the rental market, as a "new concept in urban living dealing with satisfying various residential needs within a single area." Accordingly, accommodation suitable for many different family types was provided in one- or two-bedroom apartment units, in either a two- or three-storey walk-up building or as a two-storey, two-bedroom unit in a semi-detached building. The building plans eliminated long corridors by having separate entrances and stairways serving four to six apartments, and each apartment extends from one side of the building to the other.

Coverage of the site is only about a quarter of the whole area, and the buildings are grouped to ensure plenty of light and air in the units and no "near facing" windows. The last few buildings to be built were enlarged slightly in plan by eliminating the balconies of the units, and minor changes were made to the exterior detailing. Durable materials were used throughout and, 60 years later, the buildings are fully occupied with a waiting list. The landscape is as inviting as ever, but clearly the lollypop luminaires, concrete planters, and scattered garden statuary are not original.

[Architectural drawings for The Garden Court Apartments are in the Page and Steele Collection at the Archives of Ontario. Five original drawings for the landscape survive in the Dunington-Grubb/Stensson Collection at the University of Guelph. The regular spacing of weeping trees is characteristic of the firm, as is the extensive use of hedging, especially Alpine currant (recommended in the 1939 Sheridan Nurseries catalogue as "the most satisfactory shrub for a deciduous hedge") and Japanese yew ("the best shrub available for an evergreen hedge of moderate height"). The clipped yews punctuating the hedges are original elements, but the Colorado spruce interrupting the axial view is probably a later addition.]

William N. Greer and Pleasance Kaufman Crawford

Leaside

Laid out in 1912 by Frederick G. Todd of Montreal, Leaside occupies 1,000 acres of fairly flat land between Bayview Avenue and Leslie Street, extending a mile and a half from Moore Avenue to north of Eglinton Avenue. Todd, a former student of F.L. Olmstead, is said to have been Canada's first resident landscape architect. His plan for Leaside, commissioned by the Canadian Northern Railway, is one of his less well-known schemes. Other cities and towns for which he prepared plans include Ottawa (1903), Port Mann, now part of Surrey, B.C. (1910), and the Town of Mount Royal, Quebec. (1911). Among the well-known parks that were developed to his designs are Wascana Centre, Regina (1907), the Plains of Abraham, Quebec City (1908), and Assiniboine Park, Winnipeg (1909).

Leaside was incorporated as a town in April 1913, notwithstanding it was still farmland. Within a month its provisional council had approved Todd's plan for a grid of curvilinear streets, several of which were named for people connected with the Canadian Northern Railway, including Hanna, Laird and Wicksteed. Development was hampered, however, by poor connections with the city and by the outbreak of the First World War. As part of the war effort, a munitions plant and an airfield for pilot training were established on the east side of Laird Drive, but neither initiative stimulated much house-building. Nor, it seems, did the erection of large plants for Canada Wire and Cable and the Durant Motor Car Company during the 1920s. As late as 1929, there were only 68 dwellings within the town's limits.

Leaside's fortunes changed during the following decade when almost 2,000 houses were erected there. By 1939 the population exceeded 5,000. This boom resulted in a harmony of housing design that is remarkable today. Street after street is flanked by handsome boulevard trees and tidy single-family homes in stripped-down Georgian or Tudorbethan style, each set back from the road an identical distance, on a comfortable lot with a private driveway. Although Leaside ceased to exist as a separate municipality in 1967 and many of the large factories east of Laird have been demolished more recently, the area remains a desirable neighbourhood to live in.

Stephen A. Otto

LEASIDE
PUBLIC

Don Mills

Don Mills was the first large-scale modern community of the post-war era developed solely by private enterprise. It was conceived as a garden city, as opposed to a suburb. Its planners were influenced by principles derived from Sir Ebenezer Howard's seminal work of 1898 entitled *Tomorrow: A Peaceful Path to Social Reform* (changed to *Garden Cities of Tomorrow* in 1902) and by such leading figures as Sir William Holford, an internationally renowned English town planner and educator who was active at University College, London, and Harvard University. Another significant influence on the planning principles of Don Mills was the work of two American town planners, Clarence S. Stein and Henry Wright, who in 1928 aspired to develop a garden city community of 7,500–10,000, called Radburn, in the Borough of Fairlawn, New Jersey.

At least five planning concepts were instilled at Don Mills that had not been implemented in Canada before. The first was the neighbourhood principle. The community of Don Mills was broken into four neighbourhood quadrants surrounding a regional shopping centre. Each quadrant contains a public school, a church, and a park.

The second concept was the principle of separating vehicles from pedestrians. The community is bisected by two major streets, Lawrence Avenue and Don Mills Road. A ring road, The Donway, separates the Town Centre from each neighbourhood quadrant. Pedestrian walkways provide easy access through parks to neighbourhood schools and the Town Centre. Neighbourhood roads were designed to slow vehicular traffic through residential areas by using cul-de-sacs, T-intersections and winding roads. This principle, although clearly derived from Stein's and Wright's plan for Radburn, was fully implemented in Don Mills 20 years later, when the time was ripe for the 'town for the motor age.'

The third concept was the promotion of Modern architecture and the Modern aesthetic. Don Mills Development controlled the architectural design, colours, and materials of all buildings in Don Mills. Furthermore, the corporation insisted that builders use company-approved architects – younger architects like Henry Fliess, James Murray, Irving Grossman, Michael Bach, and John B. Parkin Associates, who had been educated according to Bauhaus principles – to avert any chance of the project deteriorating into a typical post-war subdivision of builders' houses.

The fourth concept was the development of a greenbelt linked to a system of neighbourhood parks that would preserve the beauty of the surrounding ravines. The distinction between public and private ownership of lands is a crucial ingredient to the success of the greenbelt concept. Certain areas such as Wilket Creek Park were purchased by the Toronto Conservation Authority in the 1950s and became significant public amenities, kept in a natural state, while other green spaces such as E.P. Taylor's Windfields farms were sold off to builders for suburban housing developments.

The integration of industry into the community was the fifth original element in Don Mills' inception. Following the ideals of Howard's garden city, the planners of Don Mills felt it was important to include the opportunity for residents to live and work in the same satellite town so that Don Mills did not become a dormitory suburb. A sizeable component of intense residential densities – rental row houses and low-rise apartments – was essential if the town were to attract a cross-section of residents working in local industries.

Don Mills became the testing ground for many new and innovative developments in housing. Don Mills Development recognized the need for both rental housing units and single-family dwellings. James Murray's and Henry Fliess's *South Hills Village* (1956) is modeled on Pittsburgh's Chatham Village (Stein and Wright, site planners and consultant architects; Ingraham and Boyd Architects, 1932). The 190 rental units, mostly two-storey with a split-level entrance are simply built and their straightforward use of materials, such as brick and wood, is direct and honest.

As the density increased from single-family dwellings to row houses, more land was allocated for common use. South Hills Village offered private, individual green spaces and extensive communal areas facing visually pleasing internal streets. The common service facilities necessary for a communal housing development were amply and intelligently accommodated.

During the next few years in Don Mills, other new medium-density housing developments – Greenbelt Heights Village (Belcourt and Blair, 1958), Flemington Village (Irving Grossman, 1965), Yorkwoods Village Phases I and II (Klein and Sears, 1963 and 1965) – used South Hills Village as a model because it addressed the issues of human scale in both public and private spaces, devised innovative building types such as the split-level row house, and created a smaller, identifiable neighbourhood within a larger community.

In the 1950s, South Hill Village was one of three developments in Don Mills to receive a Massey Medal for distinction in architectural design.

Adapted from 'Don Mills, New Town' by Brigitte Shim, in Bureau of Architecture and Urbanism, Toronto Modern Architecture 1945–1965. *Toronto: Coach House Press, 1987, pp. 32–51.*

Brigitte Shim

The "Let's Build Program"

The first four City-owned sites

The City's new "Housing First Policy" produced the Let's Build Program and the Capital Revolving Fund. These initiatives are directed at the development of affordable housing on City-owned land and on land and in buildings owned by the private and institutional sectors.

The City, on an ongoing basis, will make parcels of its own surplus land available for the development of affordable housing. The first four parcels of surplus land were made available through a "Let's Build" proposal call issued in the fall of 1999. That call elicited almost 60 responses, of which 23 were directed at the four sites. Through a detailed review process, the list was reduced to nine and, by April 2000, four preferred proponents were isolated for more detailed discussion. The intention is to arrive at a contractual arrangement acceptable to the City and the proponents that would allow development to proceed on site. The first four preferred proponents are all, in one form or another, led by a strong, community-based non-profit organization. Each was able to convince a selection committee that they had the necessary characteristics to ensure development success. The projects range from a Single Room Occupancy project that will house mostly low-income singles, to a family-oriented townhouse/apartment structure.

419–425 Coxwell Avenue

This site was acquired by the City in 1988 as part of a Section 37 Agreement under the *Planning Act*. It is situated on the east side of Coxwell Avenue, north of Gerrard Street East.

The industrial building that originally occupied 425 Coxwell has been demolished, except for a small portion used to contain PCBs. Any redevelopment of the site will have to retain them in approved storage or relocate them in an approved manner. The building at 419 Coxwell remains in place. A fire occurred in this building in late 1991.

A 90-unit social housing development had received municipal and Ontario Municipal Board approval in 1991, but the project was aborted by the provincial government's cancellation of the non-profit housing program in 1995. It had provided for new townhouses on the northerly part of the site and 30 loft-style units at 419. Upon cancellation of the housing program, Toronto City Council placed the northerly 20m of the site under the jurisdiction of the Parks Department, thereby reducing the site's frontage to approximately 71.5m. With a depth of about 74m, the site area available for development is about 5,297m^2.

A variety of land uses surround this site. Immediately to the north is a strip of land owned by the Separate School Board, used for access to Georges-Étienne Cartier school to the east of the site. The design of the 20m-strip of

425 Coxwell, which was designated for the City's Parks Division, will take into account the School Board land by providing a more generous and attractive approach to the school from the west in addition to enhancing residential development on the subject lands. A property further to the north abuts the main east-west CN railway corridor.

The underlying zoning permits a development density of 0.6 times the area of the site. The Official Plan permits development up to one times coverage, which would allow the development of 53 units, assuming an average gross floor area of $100m^2$ per unit. As an existing building is in place on the 419 site, its reuse as contemplated in the approved site-specific zoning could yield 30 units. The remaining vacant land of the 425 Coxwell site has the potential for an additional 31 units at the density permitted in the Official Plan.

The presence of the existing building on the 419 Coxwell site offers an interesting opportunity for reuse of the shell. The previously approved development suggests one possible approach to it, using the rear of the first level for parking, with the bulk of the building being renovated to provide loft-style apartments.

2350 Finch Avenue West

This site was the former location of a firehall. It is situated on the north side of Finch Avenue West, just west of Weston Road. It has an area of $2,127m^2$ with 45.7m of frontage on Finch, and a depth of about 44.7m.

A revitalization study of the area to bring about improvements benefiting both residential and business users is underway. The preferred proponent has agreed to work as part of this ongoing process. The developer that will provide new affordable housing on this site could make a positive contribution to the re-urbanization of the area. To the extent that a housing development will be a change of use requiring planning approval, participation in this revitalization study process will be necessary to achieve the required rezoning and/or Official Plan Amendment. The redevelopment of this City-owned site could initiate an incremental process of change to the northwest corner of the Finch/Weston Road intersection while the larger changes to the southeast are gearing up.

The Official Plan designation is Industrial – Subcentre, and the zoning designation is mixed industrial under the former North York bylaw. A change will be required to accommodate a residential use.

While objectives will be derived in part from the neighbourhood revitalization study, basic principles of urban design are expected to be observed: The building should be parallel to the street to establish a street edge and to define the public realm; any surface parking should be behind the building; and the ground floor should contain uses to animate the street.

1978 Lakeshore Boulevard West

The site is situated at the northwest corner of Lakeshore Boulevard and Windermere. Its area is about 1,193.8m². The City acquired the property from the Toronto Harbour Commissioners in 1995. It is currently occupied by an Olco gas station, under a lease assigned by Joy Realty Ltd, which expired in November 1999.

The white brick, turreted gas station was erected in 1937, and designated a historic site by Bylaw 415-89. It is the only surviving Joy Oil Service Station in the City. The two structures should be preserved and reused within any proposed development. Heritage Toronto has requested that a developer enter into a Heritage Easement Agreement with the City.

The site is currently zoned to permit one times residential coverage, and a total of two times on the site. An L-shaped, single aspect structure to the north of the two gas stations would protect the site and the historic structures from road and rail noise.

The Maria Shchuka Library site

The existing Maria A. Shchuka District Library is located on the southeast corner of Eglinton Avenue West and Northcliffe Boulevard. The Eglinton frontage is a "main street" of commercial and residential uses; Northcliffe is residential in a mix of forms, from single detached to multi-storey apartments.

The existing site area is about 2,453 m², including surface parking and two existing houses owned by the library on Northcliffe Boulevard, south of the lane. The total permitted building gross floor area is 2.5 times the site area or about 6,131 m² on the entire land holding. Given a building area for the library and the Youth Resource Centre totalling 2,555 m², there remains 3,577 gross m², which can be devoted to housing. A separate housing building can be constructed, south of the library, with its address and entry off Northcliffe Boulevard.

In the Official Plan, the northern portion of the site is designated mixed use and the southern portion is high density residential. Both portions provide for up to 2.5 times coverage and a height limit of eight storeys or 24m. Each residential unit must be provided with 2.0m² of amenity space.

Any housing component on this site should be provided in a small but cost-efficient rental apartment building lining Northcliffe Boulevard. The perceived need is for "youth-oriented" housing, an appropriate match for the library and resource centre to the north. The owner of any housing to be built on the site should have strong connections to the local community and a good understanding of its housing and social needs. Coordination with the Library Board and its architect (A. J. Diamond, Donald Schmitt & Co.) on the overall development concept will be required.

Mark Guslits

The next generation of affordable housing in Toronto

The City of Toronto, in a concerted effort to create a supply of new, affordable housing, is establishing a new, more receptive and supportive environment for collaboration between the public, private, and non-profit sectors. The City has been seeking innovative expressions of interest for a number of demonstration projects to create rental and home-ownership opportunities for low- and moderate-income households. These initiatives are intended to occur on surplus, City-owned sites, as well as on land or in buildings currently owned by institutional and private sector proponents. The intent is to encourage the development of affordable housing that does not involve any direct City operating subsidies and that makes homes available to households that, until now, have been excluded from the market or forced to spend too much of their income on housing.

Very little truly affordable (or rental) housing has been produced in the 1990s in Toronto. Efforts to encourage the private sector to step in where governments left off have yielded virtually nothing in the way of "on-the-ground" housing. With this in mind, and seeing little evidence of immediate involvement from other levels of government, the City decided to make land and other special incentives available to help the non-profit and private sectors to develop rental and affordable ownership homes. In addition, it is intended that some of this new housing be targeted to specific needs groups. (Unfortunately, one current impediment to the private sector's immediate involvement is that portion of the *Municipal Act* (s. 111) that prohibits the City from directly or indirectly assisting any commercial enterprise "through the granting of bonuses in aid thereof." The removal of this impediment is being discussed between the Province and the City).

Toronto's objectives for affordable housing are to:

- produce rental units at rent levels affordable to households that otherwise cannot afford housing in the market;
- produce home-ownership units that are affordable and marketed to modest-income households who are first-time buyers and will occupy their homes;
- receive the highest land price consistent with their affordability goals and with the physical plan objectives for the sites that may be from time to time declared surplus to the City's needs;
- demonstrate the impact on affordability of any City capital cost write-downs that may be provided by the City, subject to the current impediment inherent in the *Municipal Act* prohibition against directly or indirectly granting bonuses to any commercial enterprise;
- create replicable models for affordable rental and home-ownership proj-

ects that do not require any extraordinary long-term enforcement or involvement by the City;

- demonstrate positive architectural, urban design, and planning values.

Research suggests that a combination of tools, such as reduced land costs and equalizing property tax rates for new rental buildings with the residential rate, can make it feasible to produce rental housing. Additional equity contributions or reduced financing costs could also help produce housing that is more affordable and below market prices. In response to the Final Report of the Mayor's Homelessness Action Task Force the City has instituted a "Housing First Policy" and a $10.9-million Capital Revolving Fund for Affordable Housing (CRF) to spur the development of affordable housing projects and provide limited financial assistance, where appropriate. The resources of the CRF will assist in the creation of new, affordable homes in various ways. Assistance may take the form of capital grants, loans, or forgivable loans (but not operating or service subsidy funding). Affordable ownership projects are eligible for repayable loans. As a rule of thumb, assistance from the CRF is restricted to no more than 15% to 25% of total project capital costs.

Capital assistance may be provided in the form of a reduced land price by way of a long-term lease at nominal cost, grants secured by a performance or operating agreement, or low-cost second mortgages.

The definition of "affordable housing" to apply to the CRF is to be established by Toronto City Council, according to CRF management guidelines adopted by Council. No firm definition has yet been approved. For current purposes, the definitions below are being used.

Rental housing: Affordable rents must be at or below 90% of the average market rents as measured by the 1998 Canada Mortgage and Housing Corporation rental survey, and shown in Table 1 as "maximum affordable rent." To receive funds from the CRF or other possible benefits from the City, market rent units may form part of a development, but some units (at least 25%) must fall at or below the "maximum affordable rent" levels. It is expected that rental housing meeting the Table 1 "maximum affordable rents" will meet the affordable housing definitions in the current Official Plans of the municipalities of the former Metropolitan Toronto.

Table 1: Rent levels

	Target rent (based on shelter component)	Maximum affordable rent
Bachelor/studio	$325	$530
1-bedroom	$515	$650
2-bedroom	$515–$602	$800
3-bedroom	$602–$649	$960

Home-ownership housing: The City's target market for non-rental housing would be first-time buyers who will occupy their units. Affordable housing costs would include mortgage payments, property taxes, maintenance fees in the case of condo-type units, utilities, and property insurance. Affordable housing costs should not be greater than 30% of gross household income with a 5% downpayment. Assumed interest is 7.5% with a 25-year amortization.

The City's approach to achieving its housing goals is "to create an environment in which the private sector and community providers will be willing and able to develop affordable housing for people with a range of housing needs that are not currently being met in the market." The hope is that the various mechanisms outlined above will allow the City to achieve its objectives.

Mark Guslits

Lawrence Heights: a CMHC model neighbourhood

Lawrence Avenue and Allen Road
Completed 1955–59

Lawrence Heights was designed and built between 1955 and 1959 as a model neighbourhood. It is one of four designed and built in the 1950s by the Canada Mortgage and Housing Commission (CMHC) to provide inexpensive rental accommodation at a time when Toronto faced a severe affordable-housing shortage. The neighbourhood was designed under the direction of Ian MacLennan as CMHC's Chief Architect. Building economies, and modern ideas on neighbourhood planning and design, helped give it its unique form. The plan was contemporary with Don Mills and was a startling alternative to the traditional grid of public streets and haphazard park and open space patterns that dominated Toronto at the time. Although flawed in implementation, the design is a significant example of modern planning and urban design in Canada.

In 1929, American planner Clarence Perry published *The Neighbourhood Unit: A Scheme of Arrangement for a Family Life Community*. This work was done with the Regional Planning Association, whose members included Clarence Stein, Henry Wright, and Thomas Adams. They advocated building "neighbourhoods" as the basis for city growth. Their work was widely accepted in progressive planning and design circles. It influenced the Canadian planner and educator Humphrey Carver, author of *Housing for Canadians*, and CMHC's policy and design. The ideas are the root of the design for Lawrence Heights.

Clarence Perry's six principles of neighbourhood building are summarized below and their application in the plan of Lawrence Heights follows:

1 *The size of a residential neighbourhood should be determined by the population needed for one elementary school: about 750 to 1,500 families on 150 to 300 acres.*

CMHC owned the 127-acre site at the edge of Toronto. The planned density was about 10 units to the acre, typical of north Toronto suburban residential development in the 1940s, with a total contemplated density of 1,060 people.

2. *The neighbourhood should be bounded by wide arterial roads that eliminate through traffic to the neighbourhood.*

All of the land was north of Lawrence Avenue West and west of Bathurst Street, both 30-metre-wide arterial roads. All land within the arterial block was designed for residential use. A large Sears store, the CMHC headquarters, and Bathurst Heights High School were placed on Lawrence Avenue. Initial plans for a neighbourhood of cruciform high-rise apartments designed by George Wigglesworth were rejected as inappropriate due to the proximity of the Downsview Airport and because they would have been too expensive to meet low rental-fee requirements. The cruciform high-rise apartments were replaced by a mix of walk-up apartments, townhouses, and individual houses in 1957.

3 *Within the neighbourhood there should be a hierarchy of streets, each designed to minimum widths and laid out to discourage through traffic.*

The road plan was the result of a combination of new ideas on housing and the need to keep development costs low. Varna Drive and Flemington Road together form a large, curved, gourd-shaped super block that is the defining element of the neighbourhood plan. It is about 600 metres by 450 metres, or nearly 100 acres. Varna and Flemington serve a series of apartment courts and connected open spaces within the block. Houses and townhouses were located outside of the super block. These two roads are about one quarter of the length of road required for a typical grid of streets with houses and represented considerable savings in development costs. Blossomfield Drive, Varna Road, and Replin Road connect the neighbourhood to Lawrence Avenue. All are local roads with 8.5 metres of pavement and 1.5-metre-wide concrete sidewalks. Replin Road was designed as a ceremonial boulevard from Lawrence Avenue past the CMHC headquarters to the public school. Ridgevale Drive, Rondale Boulevard, and Kirkland Boulevard were designed to connect with the neighbourhood to the east, but because they were never connected, they contribute to the social and physical isolation of the area.

4 *Streets and open spaces should make up at least 40% of any neighbourhood.*

At Lawrence Heights the connected series of open spaces in the super block established addresses for most of the units and provided pedestrian access to the elementary school without crossing a road. Apartments were grouped within the super block, their fronts facing the street and open spaces. Cars and parking were located within each group in a series of shared, at-grade parking courts. This was a local interpretation of design ideas first seen in England in Hampstead Garden Suburb and in the United States at Radburn. The idea of a public open space that was not a street did not live long. By 1983 the City assumed responsibility for the parking courts and gave them official names, which became the address system for the area, thereby inverting the front-back relationships of the original design.

The original 1955 super-block design was not altered in 1957 when an 80-metre right of way was granted for the future Spadina Freeway (now called the W.R. Allan Road and Spadina subway), even though it cut the super block in two. The expressway and subway were trenched 15 metres below the level of the neighbourhood. Flemington Drive was linked with a bridge, but the open-space system within the neighbourhood was disconnected by the trench and is currently lined with four-metre-high, concrete and steel acoustic barriers. The expressway and subway have essentially destroyed the neighbourhood super block diagram. And Replin Avenue, with its ceremonial design and institutional connections, was cut off from Lawrence Avenue because it was too close to the on-ramps of the freeway.

5 *Schools and other institutions should be grouped at a central point in the neighbourhood.*

The Lawrence Heights elementary school and its yard are the focus of the central super block. Residents within the block had access to the school without crossing public roads and those outside of the super block had only one local road to cross.

6 *Shopping areas adequate for the population should be set up at the edges of the neighbourhood, adjacent to arterial traffic.*

Local shopping was planned at the Lawrence Plaza, one of Toronto's first shopping plazas, located 400 metres east of the neighbourhood at the intersection of Lawrence Avenue and Bathurst Street. Space for shops was built inside the super block on Flemington Road, adjacent to the school. This was contrary to Clarence Perry's rules and the traditional location for stores on arterial roads. The shops have had a history of financial trouble.

The future

The neighbourhood is in public ownership and has had a reputation for social problems. The buildings are now 40 to 45 years old and require significant investment to bring them up to current standards. Redevelopment has begun at the edges of the property. Canada Lands have entered into an agreement to redevelop the CMHC property on Lawrence Avenue. This proposal is for townhouses and a 6- to 10-storey apartment at densities significantly above those of the neighbourhood. Recent downloading of federal and provincial housing interests to the municipality will make any redevelopment significantly easier to administer.

Like Regent Park, Lawrence Heights is considered to be ripe for redevelopment, but in view of the post-1970s criticisms of these early modern neighbourhoods, any redevelopment will likely feature mixed-tenure, street-oriented buildings rather than the open-space model. If capital funding, replacement housing, and resettlement issues can be resolved, this neighbourhood will probably be transformed.

The unique form of the neighbourhood is not properly understood, partly because it is badly compromised by the freeway and by the disconnection from the area to the east, and because it is threatened by redevelopment. Redevelopment of other neighbourhoods of this type, such as the Curie Barracks in Calgary, replaced modern-open space ideas with traditional streets. The buildings of Lawrence Heights may be inexpensive and uninteresting, but the unique plan and open-space character of the neighbourhood is significant.

Leo deSorcy

LEILA LANE
CATHER CRESCENT
VARNA
DRIVE
STOCKTON ROAD
FLEMINGTON
ROAD
LAWRENCE AVENUE WEST

CPD/CT

Contributors

Thanks to our contributors

As this guide accompanies you on your explorations of the city, different authors will offer their perspectives for you to consider at each stop along the way. We owe these contributors our sincere and enthusiastic thanks – for the entries they've written and the steps they've taken to make the City of Toronto a better and more interesting place to live.

Nancy Byrtus, Mark Fram, Michael McClelland

Kathryn Anderson, Preservation Officer, Culture Division, City of Toronto
Tamara Anson-Cartwright, Ontario Ministry of Citizenship, Culture and Recreation
Bob Barnett, Architect, R.E. Barnett Architect
J. Blumenson, Heritage Preservation Services, City of Toronto
Bill Bosworth, Manager, Policy and Planning, Toronto Housing Co. Inc.
Wins Bridgman, Principal, DAPR Architecture (Office for Design + Architecture + Preservation + Research)
Jon Caulfield, Associate Professor; Coordinator, Urban Studies Program, Division of Social Science, York University
Ian Chodikoff, Intern Architect, Brisbon Brook Beynon Architects
Pleasance Kaufman Crawford, Landscape Design Historian
Joan E. Crosbie, Curator, Casa Loma
Kelly Crossman, Associate Professor, Carleton University
Cathy Crowe, Street Nurse, Queen West Community Health Centre
David Dennis, Director of Urban Design, Davies Smith Development Inc.
Leo deSorcy, Urban Design Coordinator, North District, City of Toronto
Klaus Dunker, Professor, Faculty of Architecture, Landscape and Design, University of Toronto
Kathy Farrell, Secretary, Cabbagetown Preservation Association
Sean C. Fraser, Preservation Officer, Architecture, City of Toronto
Sally Gibson, Archivist, City of Toronto Archives
Robert Glover, Director, Urban Design, City of Toronto
Brad Golden and Lynne Eichenberg
William N. Greer, Architect/Heritage Consultant
Mark Guslits, Special Advisor, Housing Development
Alison Guyton, Executive Director, Habitat Services
Siamak Hariri, Partner, Taylor Hariri Pontarini Architects
Kenneth L. Hayes, Adjunct Assistant Professor, Faculty of Architecture, Landscape and Design, University of Toronto
Denis Heroux, Ontario Heritage Foundation
Robert G. Hill, Architect
Marsha Kelmans, ERA Architects Inc
Leslie M. Klein, Principal, Quadrangle Architects Limited

Bronwyn Krog, Whittington Properties Limited, Senior Director, Planning and Development
Mark Laird, Adjunct Associate Professor, Faculty of Architecture, Landscape and Design, University of Toronto
Jim Lemon, Professor, Department of Geography, University of Toronto
Janna Levitt, Levitt Goodman Architects Ltd.
Frank Lewinberg, Partner, Urban Strategies Inc.
Jerome Markson, Architect/Principal, Markson Borooah Hodgson Architects Ltd
Judy Matthews, University of Toronto, Open Space-Planning and Development
Richard Milgrom, Faculty of Environmental Studies, York University
Catherine Nasmith, Architect
Paul Northgrave, Principal, Northgrave Architect Inc.
Sylvia Novac, Research Consultant, Sylvia Novac and Associates
David Oleson, Principal, Oleson Worland Architects
John Ota
Stephen A. Otto, Consulting Historian
Ian Panabaker, Principal, ERA Architects Inc.
Marco Polo, Editor, Canadian Architect
Lewis Poplak, Urban Designer, Urban Strategies Inc.
Paul Reuber Architect, President, Paul Reuber Incorporated
Larry Wayne Richards, Dean, Faculty of Architecture, Landscape and Design, University of Toronto
Jennifer Rieger, Curatorial Assistant, Art Gallery of Ontario
Hélène St. Jacques, Queen East Business Associates
Joey Schwartz, former Board Member, Campus Co-op
John Sewell
Michael Shapcott, Manager of Government Relations and Communications, Ontario Region, Cooperative Housing Federation of Canada
Brigitte Shim, Associate Professor, Faculty of Architecture, Landscape and Design, University of Toronto
Adam Sobolak, SSAC Member and University of Toronto Fine Arts Graduate
Jeff Stinson, Jeffery Stinson Architect
Jim Ward
Cynthia Wilkey, Lawyer, Green & Chercover
Leslie E. Woo, Principal, Urban Environments
Douglas Young, Douglas Young Architect
Yvonne Sze-Man Yuen, Coordinator, Safe Neighbourhood Design Workshop

National Historic Sites in Toronto

From: Parks Canada. *Register of Designations of National Historic Significance: Commemorating Canada's History*. March 1999.

Annesley Hall
Balmoral Fire Hall
Bank of Upper Canada Building
Bead Hill
Birkbeck Building
Eaton's 7th Floor Auditorium and
  Round Room
Eglinton Theatre
Elgin and Winter Garden Theatres
Fort York
Fourth York Post Office
George Brown House
Gooderham and Worts Distillery
CNE Gouinlock Buildings/Early
  Exhibition Buildings
  Fire Hall/ Police Station
  Government Building
  Horticulture Building
  Music Building
  Press Building
  HMCS Haida
John Street CPR Roundhouse
Massey Hall
Metallic Roofing Company Offices
Montgomery's Tavern
Old Toronto City Hall & York County
  Court House
Old Toronto Post Office/Old Bank of
  Canada
Osgoode Hall
Royal Alexandra Theatre
Royal Conservatory of Music
St. Anne's Anglican Church
St. James-the-Less Anglican Church
St. Lawrence Hall
Stanley Barracks/New Fort
The Grange
Toronto Island Airport Terminal
  Building
Union Station (Canadian Pacific &
  Grand Trunk)
University College
Women's College Hospital

Parks Canada   Parcs Canada

Canada

TSA

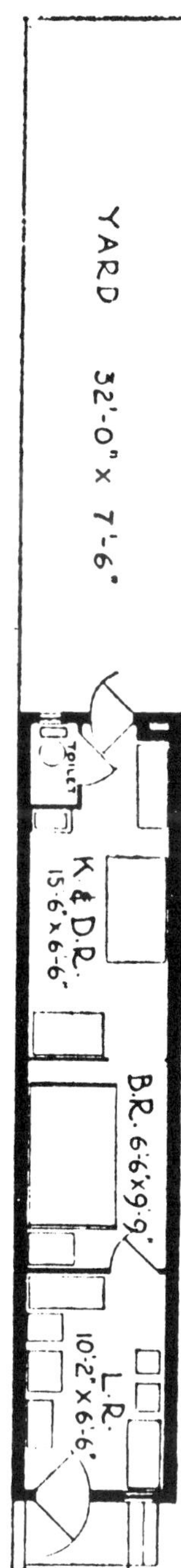

The editors

Nancy Byrtus
ERA Architects Inc

Nancy Byrtus holds a Master of Arts degree in Heritage Conservation. A former Director of the Framingham Historical Society in Massachusetts and member of the board of Heritage Vancouver, Nancy now works at ERA Architects Inc. in Toronto. She is a member of CAPHC and ICOMOS Canada.

Mark Fram
Polymath&Thaumaturge Inc

Mark Fram has written and produced many publications on architecture and heritage, notably *Well-Preserved* (Erin: Boston Mills Press, 1988/1992). He organized the SSAC's previous Toronto conference in 1987, and was President of the Society from 1988 to 1992. Mark is an award-winning graphic designer and planning consultant, and currently an adjunct assistant professor and a PhD candidate at the University of Toronto.

Michael McClelland
Principal, ERA Architects Inc

Michael McClelland is a principal of ERA Architects Inc. He is the past chair of the Toronto Society of Architects and a member of the Society for the Study of Architecture in Canada. In 1999 Michael was awarded a certificate of recognition by the Ontario Association of Architects and the Toronto Society of Architects for his contribution to the built environment.

The book

Design

Mark Fram.

Layout, cartography, and production

Typeset in Scala, Scala Sans, and Griffith Gothic. Maps and pages prepared by Mark Fram, with the invaluable aid of Stan Bevington and Rick/Simon.

Printing and binding

Coach House on bpNichol Lane.

Photography commissioned for the book

Eric J. Rogers (EJR), Mark Fram (MWF), Scott A.Weir (SAW).

Digital perspective corrections, and unannotated photographs at the beginning and end of the book (except for the photograph facing this page [EJR]), are by Mark Fram.

Project team

ERA Architects Inc.

Mark Archibald	Liyen Kan	Ian Panabaker
Maeve Burke	Marsha Kelmans	Edwin Rowse
Nancy Byrtus	Michael McClelland	Scott A. Weir
Martine Haferstroh	Curtis Murphy	David Winterton

Ken Hayes, Adjunct Professor, Faculty of Architecture, Landscape, and Design, University of Toronto.

Richard Milgrom, Faculty of Environmental Studies, York University.

Douglas Young, Douglas Young Architect.

Special thanks

Elizabeth Driver for sharing her editorial expertise and valuable insights.

Stephen A. Otto, for his expert advice and generous contribution to all facets of the project.

40

WEST EAST